Compiled by
C. Weingartner

Head, Machine Shop Department (Retired)
Ranken Technical Institute
Saint Louis, Missouri

Y0-AAG-170

machinists'

ready

reference

New Revised and Enlarged Edition

Prakken Publications — Ann Arbor, Michigan

Copyright 1977 by Prakken Publications, Inc.
All rights reserved
Printed in the United States of America
Library of Congress Catalog Card Number: 77-85648
ISBN: 0-911168-37-0

Previous editions copyrighted 1956, 1962, 1966, 1974
Additional printings 1968, 1971, 1972, 1975, 1978, 1979

HOW TO USE THIS BOOK

Convenience was the watchword in the organization of the material in this book. Related information has been put together in the same section while a comprehensive table of contents and a carefully prepared index with much cross-indexing have been designed to make information quickly obtainable.

The Index on page 215 is the best way to locate information without having to page back and forth within a section. It is arranged alphabetically, of course, and is cross-indexed extensively so that you will be able to find what you are looking for under the key word in the information you are seeking. For example, if you are seeking information on helical gears, you will find page references under both "Helical gears" and under "Gearing."

You will note that there are nine sections to this book. A brief description of each of the sections follows later in this statement of "How to Use This Book."

On each of the pages you will find a thumb-index box giving the number and title of the section. The box for the first section is near the top of the page. The box for the second section is lowered somewhat, as is the box for each succeeding section. The box for Section IX — Metric Information (a new section added in the 1974 edition) is near the top of the page. By gently folding the pages, the boxes will be exposed so you can quickly find the section you are seeking. After you have used the *Machinists' Ready Reference* enough to become acquainted with its contents, these boxes will enable you to find information very quickly.

To the Instructor

The instructor of machine-shop students or machinist apprentices should devote a portion of the teaching period to training his students in the use of this book. Call this instruction what you will—shop theory, technical information, trade principles, or practical shop mathematics. Remember, this book is not an instruction book, but a reference book. The person must learn how to use it. As the instructor, you should develop practical problems and questions to illustrate the value and application of the material in this book.

To the Student or Trainee

Make every effort to understand completely every phase of the work-job assigned to you. If you are operating a standard machine tool, check to see that the speeds and feeds are set according to recommendations. Cutting tool shapes are most important for efficient cutting on machines. Mathematical problems must often be solved in order to set up the machine properly. When in doubt about any detail, consult with your instructor or lead man but look up the data you need first.

To the Journeyman Machinist or Toolmaker

As a first-class mechanic, you realize that it is impractical if not impossible to remember all of the detailed information used in your work. You already know why the information is necessary and how to use it. For you, the important thing to know is where to find the information in this convenient data book so you can have it available as you need it.

The following will brief you on what to look for in each section:

Section I — Mathematical Information

The section on mathematical information contains most of the data, rules, and formulas necessary to solve the mathematical problems encountered in the machinist and toolmaking trades. Higher mathematics is not involved in this work, but a fundamental knowledge of arithmetic is essential. Some mechanics may tell you that it is not necessary to be able to work these problems, but this is not so. Every time you use a rule to measure, divide a dimension on a drawing, or change a fraction to a decimal, some mathematics is involved. The more you can learn to use this section, the more time you can save and the more guesswork you can eliminate.

Section II — Drills

Probably one of the most efficient cutting tools we have is the common twist drill. In order to cut most efficiently, it must be properly ground and run at the correct speed and feed for the particular material being drilled. If a cutting fluid is used, speeds can be raised and production increased. Drills come in a variety of sizes, such as fractional, numbered, lettered, and metric. They are used on a variety of machine tools. This section furnishes you much of the information required by the machinist and toolmaker on drills and drilling.

Section III — Tapers

The section on tapers is important to the machinist because he is called on to turn, plane, bore, and grind tapers on a variety of parts. You should, therefore, get acquainted with the common standard tapers—The American Standard, the Brown and Sharpe, and milling-machine spindle. How to determine the taper per inch, or percentage of taper, from a sample piece of work is a common problem. A study of the data and the rules in this section will help you handle your taper problems.

Section IV — Screw Threads

In order that parts may be assembled and disassembled, they are frequently joined together with screws. This calls for many threaded parts having external and internal threads. A good draftsman or machinist must have a wide knowledge of screw-thread standards and shapes. He must know the proper sizes to drill or bore the work to produce threads of a certain size and shape. He must understand how to accurately measure screw threads and be familiar with the proper tolerances necessary for the required fits. The section on screw threads provides you with much of this needed information.

Section V — Milling, Shaping, Turning

The machinist is called on to operate all of the standard machine tools commonly used in a machine shop such as the lathe, drill press, shaper, milling machine, and others. All of these machines call for some special data as to tool shapes, methods of sharpening the tools, cutting speeds, proper feeds, and cutter lubrication for different materials. To operate these machines efficiently, consult this section on milling, shaping, and turning for proper recommendations.

Section VI — Gears

A knowledge of gears of different types is important in the design of most machines. Although gears are produced by many different specialized machines, gears that are produced in small quantity may be made on regular milling machines in the machine shop. Basic rules and formulas for calculating the dimensions of the four most common types of gears are given in the section on gears.

Section VII — Weights, Gages, Tolerances

Dimensions to close tolerances are essential in modern machine-shop practice. This section includes data on gages and tolerances as well as charts for estimating the weights of bar or sheet steel. Stock thickness may be checked against a gage or the reverse.

The machinist is called on to make various types of fits to facilitate the proper functioning of the mechanism. Some parts are pressed together with force, while in other cases, shafts must turn freely in their bearings. All this calls for exact tolerances in the machining of parts to be assembled. Recommendations for the proper sizing of mating parts can be found in this section.

Toolmakers and inspectors use pins, discs, and balls of different diameters to check dovetails, angles, and cavities. The use of these accessories simplifies the measuring of such parts, but mathematical calculations are necessary to determine what the dimensions should be. Formulas for use in this class of work are found in this section.

Another device used for setting up or for checking angles in precision work is the sine bar. A special set of sine-bar tables gives the proper dimensions for setting a sine bar at the certain angle.

Gage blocks accurately machined to within a few millionths of an inch are furnished in sets and consist of various combinations of sizes. In setting up these blocks, the toolmaker must know the sizes available.

6

Section VIII — General Information

This section includes some machine-shop rules not found elsewhere in the book. The machinist is often called on to change from taper per inch to degrees in order to set the lathe compound, boring-mill head, or grinder head.

This section also contains some fundamental information on the heat treatment of steels. Even though the machinist is not always responsible for actually heat-treating the parts, a knowledge of the terms used and what happens when the parts are heat-treated is important. A chart for comparing the various hardness testing readings is useful in evaluating heat-treating results.

Section IX — Metric Information

In this section, information is furnished to help you make the change from the English to the Metric System. Until the final changeover is accomplished, it will be necessary to use comparison charts for dual dimensioning, comparing the readings of various measuring instruments, and interpreting metric charts and other data. A number of metric comparison charts and some information on metric threads are furnished in this section.

Now that you have had a preview, turn to the table of contents, pages 9 to 12, and read it through. Next, study each chart to get an idea of its range. When you are seeking specific information, remember to refer to the index on page 215 to help you find information quickly.

Please remember this—the more you use your book, the more familiar you will become with its contents. It is not necessary to memorize the material, but it is important that you know how and where to locate the information when you need it.

The size and shape of this book have been designed so that it can be easily carried in a shop apron pocket or kept in a tool box. The spiral binding makes it possible to open up the book to any page and it will lie flat for easy reference at your work station. The cover is specially coated so that grease and dirt marks can be readily wiped off. With proper care, the book should last for many years of useful reference.

The Author

PREFACE

This book contains a large percentage of the fundamental data needed by the machinist and allied workers. The draftsmen and designers who make ready the drawings and plans for the machinist and toolmaker, the production planner, and the inspector, all can make use of the information contained herein.

The machinist or toolmaker must constantly look up information or solve shop problems in his work. He becomes a more valuable mechanic if he can solve these problems or look up the information without help from an instructor, lead man, or foreman. Moreover, there is a great deal of self-satisfaction that comes from being able to work out a problem on your own. This book is designed to aid in this procedure.

The student of machine-shop practice, the apprentice machinist, the journeyman machinist or toolmaker, the foreman or superintendent, all will find the information in this book presented in easily found and convenient form. It is not designed to take the place of the larger, more expensive handbooks but rather to supplement them—a small inexpensive book which will be handy to use right on the job.

Permission has been graciously granted by several companies and organizations to reprint material in this book. The author wishes to express his appreciation to these companies and his gratefulness to those who offered suggestions for making this book more valuable. Acknowledgment is due to the following:

Am. Soc. of Mech. Engineers	Lufkin Rule Co.
Barber-Colman Co.	M.T.I. Corp. for Mitutoyo Mfg. Co.
Beloit Tool Corporation	Morse Twist Drill & Machine Co.
Brown & Sharpe Mfg. Co.	National Twist Drill & Tool Co.
Carboloy Company, Inc.	Niles-Bement-Pond Co.
Cincinnati Milling Machine Co.	Norton Co.
The Cincinnati Shaper Co.	Republic Steel Corp.
The Cleveland Twist Drill Co.	Sheffield Corp.
DoAll Company	Shell Oil Co.
Grinding Wheel Institute	South Bend Lathe Works
Illinois Tool Works	Standard Pressed Steel Co.
Industrial Fasteners Institute	L. S. Starrett Co.
R.K.LeBlond Machine Tool Co.	Union Tool Co.

CONTENTS

IV. Screw Threads

V. Milling, Shaping, Turning

VI. Gears

VII. Weights, Gages, Tolerances

VIII. General Information

IX. Metric Information

DECIMAL EQUIVALENT PARTS OF AN INCH

RULE: To convert a common fraction to a decimal fraction, divide the numerator by the denominator. Example: 1/4 = 1 ÷ 4 = .250. To convert a decimal fraction to a common fraction, write the decimal — without the decimal point — as the numerator and 1, followed by as many zeros as there were places in the decimal, as the denominator; then divide both numerator and denominator by the largest number that will go equally into both. Example: .250 = 250/1000 = 1/4.

In machine shop practice, practically all measurements are expressed in either fractions of an inch or their decimal equivalents. At times it is necessary to convert one to the other. To the left is the rule by which this may be done, and below is a handy table, worked out on the basis of sixty-fourths of an inch.

16	32	64	DECIMAL
		1/64	.015625
	1/32	2/64	.031250
		3/64	.046875
1/16	2/32	4/64	.062500
		5/64	.078125
	3/32	6/64	.093750
		7/64	.109375
1/8	4/32	8/64	.125000
		9/64	.140625
	5/32	10/64	.156250
		11/64	.171875
3/16	6/32	12/64	.187500
		13/64	.203125
	7/32	14/64	.218750
		15/64	.234375

16	32	64	DECIMAL
1/4	8/32	16/64	.250000
		17/64	.265625
	9/32	18/64	.281250
		19/64	.296875
5/16	10/32	20/64	.312500
		21/64	.328125
	11/32	22/64	.343750
		23/64	.359375
3/8	12/32	24/64	.375000
		25/64	.390625
	13/32	26/64	.406250
		27/64	.421875
7/16	14/32	28/64	.437500
		29/64	.453125
	15/32	30/64	.468750
		31/64	.484375

16	32	64	DECIMAL
1/2	16/32	32/64	.500000
		33/64	.515625
	17/32	34/64	.531250
		35/64	.546875
9/16	18/32	36/64	.562500
		37/64	.578125
	19/32	38/64	.593750
		39/64	.609375
5/8	20/32	40/64	.625000
		41/64	.640625
	21/32	42/64	.656250
		43/64	.671875
11/16	22/32	44/64	.687500
		45/64	.703125
	23/32	46/64	.718750
		47/64	.734375

16	32	64	DECIMAL
3/4	24/32	48/64	.750000
		49/64	.765625
	25/32	50/64	.781250
		51/64	.796875
13/16	26/32	52/64	.812500
		53/64	.828125
	27/32	54/64	.843750
		55/64	.859375
7/8	28/32	56/64	.875000
		57/64	.890625
	29/32	58/64	.906250
		59/64	.921875
15/16	30/32	60/64	.937500
		61/64	.953125
	31/32	62/64	.968750
		63/64	.984375

TABLE of DECIMAL EQUIVALENTS
of

8ths, 16ths, 32ds and 64ths of an inch

8ths

$\frac{1}{8}$ = .125
$\frac{1}{4}$ = .250
$\frac{3}{8}$ = .375
$\frac{1}{2}$ = .500
$\frac{5}{8}$ = .625
$\frac{3}{4}$ = .750
$\frac{7}{8}$ = .875

16ths

$\frac{1}{16}$ = .0625
$\frac{3}{16}$ = .1875
$\frac{5}{16}$ = .3125
$\frac{7}{16}$ = .4375
$\frac{9}{16}$ = .5625
$\frac{11}{16}$ = .6875
$\frac{13}{16}$ = .8125
$\frac{15}{16}$ = .9375

32ds

$\frac{1}{32}$ = .03125
$\frac{3}{32}$ = .09375

$\frac{5}{32}$ = .15625
$\frac{7}{32}$ = .21875
$\frac{9}{32}$ = .28125
$\frac{11}{32}$ = .34375
$\frac{13}{32}$ = .40625
$\frac{15}{32}$ = .46875
$\frac{17}{32}$ = .53125
$\frac{19}{32}$ = .59375
$\frac{21}{32}$ = .65625
$\frac{23}{32}$ = .71875
$\frac{25}{32}$ = .78125
$\frac{27}{32}$ = .84375
$\frac{29}{32}$ = .90625
$\frac{31}{32}$ = .96875

64ths

$\frac{1}{64}$ = .015625
$\frac{3}{64}$ = .046875
$\frac{5}{64}$ = .078125
$\frac{7}{64}$ = .109375
$\frac{9}{64}$ = .140625
$\frac{11}{64}$ = .171875
$\frac{13}{64}$ = .203125
$\frac{15}{64}$ = .234375

$\frac{17}{64}$ = .265625
$\frac{19}{64}$ = .296875
$\frac{21}{64}$ = .328125
$\frac{23}{64}$ = .359375
$\frac{25}{64}$ = .390625
$\frac{27}{64}$ = .421875
$\frac{29}{64}$ = .453125
$\frac{31}{64}$ = .484375
$\frac{33}{64}$ = .515625
$\frac{35}{64}$ = .546875
$\frac{37}{64}$ = .578125
$\frac{39}{64}$ = .609375
$\frac{41}{64}$ = .640625
$\frac{43}{64}$ = .671875
$\frac{45}{64}$ = .703125
$\frac{47}{64}$ = .734375
$\frac{49}{64}$ = .765625
$\frac{51}{64}$ = .796875
$\frac{53}{64}$ = .828125
$\frac{55}{64}$ = .859375
$\frac{57}{64}$ = .890625
$\frac{59}{64}$ = .921875
$\frac{61}{64}$ = .953125
$\frac{63}{64}$ = .984375

Useful Information

To find the circumference of a circle, multiply the diameter by 3.1416.

To find the diameter of a circle, multiply the circumference by .31831.

To find the area of a circle, multiply the square of the diameter by .7854.

To find the surface of a ball (sphere), multiply the square of the diameter by 3.1416.

To find the side of a square inscribed in a circle, multiply the diameter by .70711.

To find the diameter of a circle to circumscribe a square, multiply one side by 1.4142.

To find the cubic inches (volume) in a ball, multiply the cube of the diameter by .5236.

Doubling the diameter of a pipe increases its capacity four times.

The radius of a circle × 6.283185 equals the circumference.

The square of the circumference of a circle × .07958 equals the area.

Half the circumference of a circle × half its diameter equals the area.

The circumference of a circle × .159155 equals the radius.

The square root of the area of a circle × .56419 equals the radius.

The square root of the area of a circle × 1.12838 equals the diameter.

TABLE OF SQUARES AND SQUARE ROOTS

n	n^2	$\sqrt{n}$	n	n^2	$\sqrt{n}$
1	1	1.000	51	2 601	7.141
2	4	1.414	52	2 704	7.211
3	9	1.732	53	2 809	7.280
4	16	2.000	54	2 916	7.348
5	25	2.236	55	3 025	7.416
6	36	2.449	56	3 136	7.483
7	49	2.646	57	3 249	7.550
8	64	2.828	58	3 364	7.616
9	81	3.000	59	3 481	7.681
10	100	3.162	60	3 600	7.746
11	121	3.317	61	3 721	7.810
12	144	3.464	62	3 844	7.874
13	169	3.606	63	3 969	7.937
14	196	3.742	64	4 096	8.000
15	225	3.873	65	4 225	8.062
16	256	4.000	66	4 356	8.124
17	289	4.123	67	4 489	8.185
18	324	4.243	68	4 624	8.246
19	361	4.359	69	4 761	8.307
20	400	4.472	70	4 900	8.367
21	441	4.583	71	5 041	8.426
22	484	4.690	72	5 184	8.485
23	529	4.796	73	5 329	8.544
24	576	4.899	74	5 476	8.602
25	625	5.000	75	5 625	8.660
26	676	5.099	76	5 776	8.718
27	729	5.196	77	5 929	8.775
28	784	5.292	78	6 084	8.832
29	841	5.385	79	6 241	8.888
30	900	5.477	80	6 400	8.944
31	961	5.568	81	6 561	9.000
32	1 024	5.657	82	6 724	9.055
33	1 089	5.745	83	6 889	9.110
34	1 156	5.831	84	7 056	9.165
35	1 225	5.916	85	7 225	9.220
36	1 296	6.000	86	7 396	9.274
37	1 369	6.083	87	7 569	9.327
38	1 444	6.164	88	7 744	9.381
39	1 521	6.245	89	7 921	9.434
40	1 600	6.325	90	8 100	9.487
41	1 681	6.403	91	8 281	9.539
42	1 764	6.481	92	8 464	9.592
43	1 849	6.557	93	8 649	9.644
44	1 936	6.633	94	8 836	9.695
45	2 025	6.708	95	9 025	9.747
46	2 116	6.782	96	9 216	9.798
47	2 209	6.856	97	9 409	9.849
48	2 304	6.928	98	9 604	9.899
49	2 401	7.000	99	9 801	9.950
50	2 500	7.071	100	10 000	10.000

Angle at Center and Length of Chord for Spacing off the Circumference of a Circle

Spaces on the Circle	Angle at Center		*Length of Chord
	Deg.	Min.	
3	120	..	0:866025
4	90	..	0.707106
5	72	..	0.587785
6	60	..	0.500000
7	51	26	0.433883
8	45	..	0.382683
9	40	..	0.342020
10	36	..	0.309017
11	32	44	0.281732
12	30	..	0.258819
13	27	41	0.239315
14	25	43	0.222520
15	24	..	0.207911
16	22	30	0.195090
17	21	11	0.183749
18	20	..	0.173648
19	18	57	0.164594
20	18	..	0.156434
21	17	9	0.149042
22	16	22	0.142314
23	15	39	0.136166
24	15	..	0.130526
25	14	24	0.125333
26	13	51	0.120536
27	13	20	0.116092
28	12	51	0.111964
29	12	25	0.108118
30	12	..	0.104528
31	11	37	0.101168
32	11	15	0.098017
33	10	55	0.095056
34	10	35	0.092268
35	10	17	0.089639
36	10	..	0.087155
37	9	44	0.084805
38	9	28	0.082579
39	9	14	0.080466
40	9	..	0.078459

*NOTE: This table is calculated for circles having a diameter equal to 1. For other circles, multiply the length given by the circle's diameter.

Coordinate Constants

Used for equally spacing holes from reference lines, as in jig-boring practice.

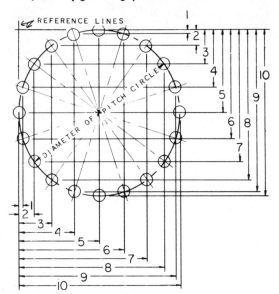

The information taken from this 20-hole layout chart may also be used for 4, 5, or 10 holes.

1. -.024472	6. -.654508
2. -.095492	7. -.793893
3. -.206107	8. -.904508
4. -.345492	9. -.975528
5. -.500000	10. 1.000000

Rule: Multiply the constants above by the diameter of the given pitch circle to obtain the actual dimensions from the reference lines.

Coordinate Constants

Used for equally spacing holes from reference lines, as in jig-boring practice.

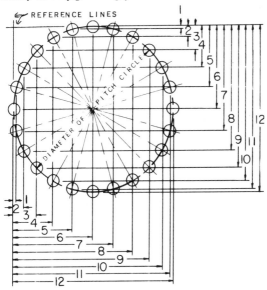

Information taken from this 24-hole layout chart may also be used for 3, 4, 6, 8, or 12 holes.

1.	-.017037	7.	-.629410
2.	-.066987	8.	-.750000
3.	-.146447	9.	-.853553
4.	-.250000	10.	-.933013
5.	-.370590	11.	-.982953
6.	-.500000	12.	1.000000

Rule: Multiply the constants above by the diameter of the given pitch circle to obtain the actual dimensions from the reference lines.

AREA OF PLANE FIGURES

| | GIVEN | SOUGHT |

Sector of Circle

Radius, r, and center angle, θ.   area $= \pi r^2 \dfrac{\theta}{360}$

Segment of Circle

Radius, r, and center angle, θ,   area $=$
$$\frac{r^2}{2}\left(\frac{\pi\theta}{180} - \sin\theta\right)$$

For center angles above 90°,   area $=$
$$\frac{r^2}{2}\left(\frac{\pi\theta}{180} - \sin(180 - \theta)\right)$$

Ellipse

Major axis, a, and minor axis, b,   area $= .78540 \times ab$

Parabola

Base, b, and height, h,   area $= \dfrac{2bh}{3}$

Irregular Plane Surface

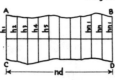

Divide any plane surface, A, B, C, D, along a line, a – b, into an even number, n, of parallel and sufficiently small strips, d, whose ordinates are h_1, h_2, h_3, h_4, h_5......h_{n-1}, h_n, h_{n+1}, and considering contours between three ordinates as parabolic curves, then for section ABCD.

$$\text{area} = \frac{d}{3}\left[h_1 + h_{n+1} + 4(h_2 + h_4 + h_6 \ldots + h_n) + 2(h_3 + h_5 + h_7 \ldots + h_{n-1})\right]$$

or, approximately, Area = Sum of ordinates x width, d.

AREA OF PLANE FIGURES

	GIVEN	SOUGHT

Triangle
Base, b, and height, h, perpendicular thereto,

$$\text{area} = \frac{bh}{2}$$

All three sides, a, b, c; then
$$s = \frac{a+b+c}{2},$$

$$\text{area} = \sqrt{s(s-a)\,(s-b)\,(s-c)}$$

Trapezium
Two heights, h and h', and resulting distances, a, b and c on base,

$$\text{area} = \frac{b(h+h') + ah + ch'}{2}$$

Data for figuring area of triangles produced by a diagonal,

area = sum of areas of both triangles

Trapezoid
Parallel sides, a and b, and height, h, perpendicular thereto,

$$\text{area} = \frac{(a+b)h}{2}$$

Parallelogram
Base, b, and height, h, perpendicular thereto,

area = bh

Regular Polygon
Side, a, and radius of inscribed circle, r,

$$\text{area} = \frac{\text{sum of sides x r}}{2}$$

Circle
Radius, r; constant $\pi = 3.141593$.
In a circle of same area as square, r = .56419 x side.
In a square of same area as circle, side = 1.77246r

area = πr^2
circumference = $2\pi r$

DISTANCE ACROSS CORNERS OF HEXAGONS AND SQUARES

$$D = 1.1547\, d$$
$$E = 1.4142\, d$$

DIMENSIONS IN INCHES

d	D	E	d	D	E	d	D	E
1/4	0.2886	0.3535	1 1/4	1.4434	1.7677	2 5/16	2.6702	3.2703
9/32	0.3247	0.3977	1 9/32	1.4794	1.8119	2 3/8	2.7424	3.3587
5/16	0.3608	0.4419	1 5/16	1.5155	1.8561	2 7/16	2.8145	3.4471
11/32	0.3968	0.4861	1 11/32	1.5516	1.9003	2 1/2	2.8867	3.5355
3/8	0.4329	0.5303	1 3/8	1.5877	1.9445	2 9/16	2.9583	3.6239
13/32	0.4690	0.5745	1 13/32	1.6238	1.9887	2 5/8	3.0311	3.7123
7/16	0.5051	0.6187	1 7/16	1.6598	2.0329	2 11/16	3.1032	3.8007
15/32	0.5412	0.6629	1 15/32	1.6959	2.0771	2 3/4	3.1754	3.8891
1/2	0.5773	0.7071	1 1/2	1.7320	2.1213	2 13/16	3.2476	3.9794
17/32	0.6133	0.7513	1 17/32	1.7681	2.1655	2 7/8	3.3197	4.0658
9/16	0.6494	0.7955	1 9/16	1.8042	2.2097	2 15/16	3.3919	4.1542
19/32	0.6855	0.8397	1 19/32	1.8403	2.2539	3	3.4641	4.2426
5/8	0.7216	0.8839	1 5/8	1.8764	2.2981	3 1/16	3.5362	4.3310
21/32	0.7576	0.9281	1 21/32	1.9124	2.3423	3 1/8	3.6084	4.4194
11/16	0.7937	0.9723	1 11/16	1.9485	2.3865	3 3/16	3.6806	4.5078
23/32	0.8298	1.0164	1 23/32	1.9846	2.4306	3 1/4	3.7527	4.5962
3/4	0.8659	1.0606	1 3/4	2.0207	2.4708	3 5/16	3.8249	4.6846
25/32	0.9020	1.1048	1 25/32	2.0568	2.5190	3 3/8	3.8971	4.7729
13/16	0.9380	1.1490	1 13/16	2.0929	2.5632	3 7/16	3.9692	4.8613
27/32	0.9741	1.1932	1 27/32	2.1289	2.6074	3 1/2	4.0414	4.9497
7/8	1.0102	1.2374	1 7/8	2.1650	2.6516	3 9/16	4.1136	5.0381
29/32	1.0463	1.2816	1 29/32	2.2011	2.6958	3 5/8	4.1857	5.1265
15/16	1.0824	1.3258	1 15/16	2.2372	2.7400	3 11/16	4.2579	5.2149
31/32	1.1184	1.3700	1 31/32	2.2733	2.7842	3 3/4	4.3301	5.3033
1	1.1547	1.4142	2	2.3094	2.8284	3 13/16	4.4023	5.3917
1 1/32	1.1907	1.4584	2 1/32	2.3453	2.8726	3 7/8	4.4744	5.4801
1 1/16	1.2268	1.5026	2 1/16	2.3815	2.9168	3 15/16	4.5466	5.5684
1 3/32	1.2629	1.5468	2 3/32	2.4176	2.9610	4	4.6188	5.6568
1 1/8	1.2990	1.5910	2 1/8	2.4537	3.0052	4 1/8	4.7631	5.8336
1 5/32	1.3351	1.6352	2 5/32	2.4898	3.0494	4 1/4	4.9074	6.0104
1 3/16	1.3712	1.6793	2 3/16	2.5259	3.0936	4 3/8	5.0518	6.1872
1 7/32	1.4073	1.7235	2 1/4	2.5981	3.1820	4 1/2	5.1961	6.3639

DISTANCE ACROSS CORNERS OF HEXAGONS AND SQUARES

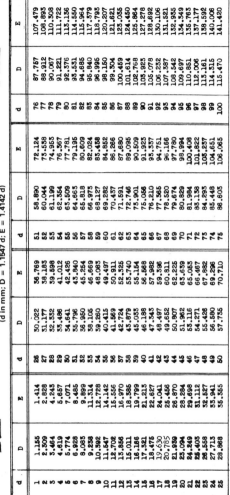

(d in mm; D = 1.1547 d; E = 1.4142 d)

d	D	E	d	D	E	d	D	E	d	D	E
1	1.155	1.414	26	30.022	36.769	51	58.890	72.124	76	87.757	107.479
2	2.309	2.828	27	31.177	38.183	52	60.044	73.558	77	88.912	108.893
3	3.464	4.243	28	32.332	39.598	53	61.199	74.955	78	90.067	110.308
4	4.619	5.657	29	33.486	41.012	54	62.354	76.367	79	91.221	111.722
5	5.774	7.071	30	34.641	42.426	55	63.509	77.781	80	92.376	113.136
6	6.928	8.485	31	35.796	43.840	56	64.665	79.195	81	93.531	114.550
7	8.083	9.899	32	36.950	45.254	57	65.818	80.609	82	94.685	115.964
8	9.238	11.314	33	38.105	46.669	58	66.975	82.024	83	95.840	117.579
9	10.392	12.728	34	39.260	48.083	59	68.127	83.458	84	96.995	118.795
10	11.547	14.142	35	40.415	49.497	60	69.282	84.852	85	98.150	120.207
11	12.702	15.556	36	41.569	50.911	61	70.437	86.266	86	99.504	121.621
12	13.856	16.970	37	42.724	52.325	62	71.591	87.680	87	100.459	123.035
13	15.011	18.386	38	43.879	53.740	63	72.746	89.095	88	101.614	124.450
14	16.166	19.799	39	45.035	55.154	64	73.901	90.509	89	102.768	125.864
15	17.321	21.213	40	46.188	56.568	65	75.056	91.925	90	103.925	127.278
16	18.475	22.627	41	47.343	57.982	66	76.210	93.337	91	105.078	128.692
17	19.630	24.041	42	48.497	59.396	67	77.365	94.751	92	106.232	130.106
18	20.785	25.456	43	49.652	60.811	68	78.520	96.166	93	107.587	131.521
19	21.939	26.870	44	50.807	62.225	69	79.674	97.780	94	108.542	132.935
20	23.094	28.284	45	51.962	63.639	70	80.829	98.994	95	109.697	134.549
21	24.249	29.698	46	53.116	65.053	71	81.984	100.408	96	110.851	135.763
22	25.403	31.112	47	54.271	66.467	72	83.138	101.822	97	112.006	137.177
23	26.558	32.527	48	55.426	67.882	73	84.293	103.237	98	113.161	138.592
24	27.713	33.941	49	56.580	69.296	74	85.448	104.651	99	114.515	140.006
25	28.868	35.355	50	57.755	70.710	75	86.603	106.065	100	115.470	141.420

TRIGONOMETRIC FUNCTIONS

Radius $1 = \sin^2 \alpha + \cos^2 \alpha = \sin \alpha \, \mathrm{cosec}\, \alpha = \cos \alpha \sec \alpha = \tan \alpha \cot \alpha$

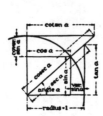

$$\sin \alpha = \frac{\cos \alpha}{\cot \alpha} = \frac{1}{\mathrm{cosec}\, \alpha}$$
$$= \cos \alpha \tan \alpha = \sqrt{1 - \cos^2 \alpha}$$

$$\cos \alpha = \frac{\sin \alpha}{\tan \alpha} = \frac{1}{\sec \alpha}$$
$$= \sin \alpha \cot \alpha = \sqrt{1 - \sin^2 \alpha}$$

$$\tan \alpha = \frac{\sin \alpha}{\cos \alpha} = \frac{1}{\cot \alpha} = \sin \alpha \sec \alpha$$

$$\cot \alpha = \frac{\cos \alpha}{\sin \alpha} = \frac{1}{\tan \alpha} = \cos \alpha \, \mathrm{cosec}\, \alpha$$

$$\sec \alpha = \frac{\tan \alpha}{\sin \alpha} = \frac{1}{\cos \alpha}$$

$$\mathrm{cosec}\, \alpha = \frac{\cot \alpha}{\cos \alpha} = \frac{1}{\sin \alpha}$$

PROPERTIES OF THE CIRCLE

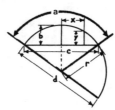

Circumference of Circle of Dia. 1 = π = 3.14159265
Circumference of Circle = $2 \pi r$
Dia. of Circle = Circumference x 0.31831
Diameter of Circle of equal periphery as Square = side x 1.27324
Side of Square of equal periphery as Circle = diameter x 0.78540
Diameter of Circle circumscribed about Square = side x 1.41421
Side of Square inscribed in Circle = diameter x 0.70711

Arc, $\quad a = \dfrac{\pi r a}{180^\circ} = 0.017453\, r\, a^\circ$ $\qquad$ Angle, $\quad a = \dfrac{180^\circ a}{\pi r} = 57.29578\, \dfrac{a}{r}$

Radius, $\quad r = \dfrac{4 b^2 + c^2}{8 b}$ $\qquad$ Diameter, $d = \dfrac{4 b^2 + c^2}{4 b}$

Rise, $\quad b = 2\sqrt{2\, b\, r - b^2} = 2r \sin\dfrac{a}{2}$ $\qquad$ Chord, $c = 2\sqrt{2\, b\, r - b^2} = 2r \sin\dfrac{a}{2}$

Rise, $\quad b = r - \tfrac{1}{2}\sqrt{4\, r^2 - c^2} = \dfrac{c}{2}\tan\dfrac{a}{4} = 2\, r \sin^2\dfrac{a}{4}$

Rise, $\quad b = r + y - \sqrt{r^2 - x^2} \qquad y = b - r + \sqrt{r^2 - x^2} \qquad x = \sqrt{r^2 - (r + y - b)^2}$

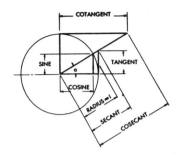

FORMULAS FOR FINDING FUNCTIONS OF ANGLES	
$\dfrac{\text{Side opposite}}{\text{Hypotenuse}}$	= SINE
$\dfrac{\text{Side adjacent}}{\text{Hypotenuse}}$	= COSINE
$\dfrac{\text{Side opposite}}{\text{Side adjacent}}$	= TANGENT
$\dfrac{\text{Side adjacent}}{\text{Side opposite}}$	= COTANGENT
$\dfrac{\text{Hypotenuse}}{\text{Side adjacent}}$	= SECANT
$\dfrac{\text{Hypotenuse}}{\text{Side opposite}}$	= COSECANT

FORMULAS FOR FINDING THE LENGTH OF SIDES FOR RIGHT-ANGLE TRIANGLES WHEN AN ANGLE AND SIDE ARE KNOWN	
Length of side opposite	Hypotenuse × Sine Hypotenuse ÷ Cosecant Side adjacent × Tangent Side adjacent ÷ Cotangent
Length of side adjacent	Hypotenuse × Cosine Hypotenuse ÷ Secant Side opposite × Cotangent Side opposite ÷ Tangent
Length of Hypotenuse	Side opposite × Cosecant Side opposite ÷ Sine Side adjacent × Secant Side adjacent ÷ Cosine

25

RIGHT TRIANGLES

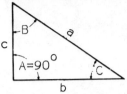

TO FIND SIDES	FORMULAS	
a	$\sqrt{b^2 + c^2}$	
a	$c \times$ Cosec. C	$\dfrac{c}{\text{sine C}}$
a	$c \times$ Seçant B	$\dfrac{c}{\text{Cosine B}}$
a	$b \times$ Cosec. B	$\dfrac{b}{\text{Sine B}}$
a	$b \times$ Secant C	$\dfrac{b}{\text{Cosine C}}$
b	$\sqrt{a^2 - c^2}$	
b	$a \times$ Sine B	$\dfrac{a}{\text{Cosecant B}}$
b	$a \times$ Cos. C	$\dfrac{a}{\text{Secant C}}$
b	$c \times$ Tan. B	$\dfrac{c}{\text{Cotangent B}}$
b	$c \times$ Cot. C	$\dfrac{c}{\text{Tangent C}}$
c	$\sqrt{a^2 - b^2}$	
c	$a \times$ Cos. B	$\dfrac{a}{\text{Secant B}}$
c	$a \times$ Sine C	$\dfrac{a}{\text{Cosecant C}}$
c	$b \times$ Cot. B	$\dfrac{b}{\text{Tangent B}}$
c	$b \times$ Tan C	$\dfrac{b}{\text{Cotangent C}}$

RIGHT TRIANGLES

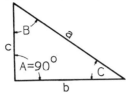

TO FIND ANGLES	FORMULAS	
C	$\dfrac{c}{a}$ = Sine C	90° − B
C	$\dfrac{b}{a}$ = Cosine C	90° − B
C	$\dfrac{c}{b}$ = Tan. C	90° − B
C	$\dfrac{b}{c}$ = Cotan. C	90° − B
C	$\dfrac{a}{b}$ = Secant C	90° − B
C	$\dfrac{a}{c}$ = Cosec. C	90° − B
B	$\dfrac{b}{a}$ = Sine B	90° − C
B	$\dfrac{c}{a}$ = Cosine B	90° − C
B	$\dfrac{b}{c}$ = Tan. B	90° − C
B	$\dfrac{c}{b}$ = Cotan. B	90° − C
B	$\dfrac{a}{c}$ = Secant B	90° − C
B	$\dfrac{a}{b}$ = Cosec. B	90° − C

OBLIQUE TRIANGLES

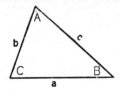

TO FIND	KNOWN	SOLUTION
C	A·B	$180° - (A + B)$
b	a·B·A	$\dfrac{a \times \text{Sin. B}}{\text{Sin. A}}$
c	a·A·C	$\dfrac{a \times \text{Sin. C}}{\text{Sin. A}}$
Tan. A	a·C·b	$\dfrac{a \times \text{Sin. C}}{b - (a \times \text{Cos. C})}$
B	A·C	$180° - (A + C)$
Sin. B	b·A·a	$\dfrac{b \times \text{Sin. A}}{a}$
A	B·C	$180° - (B + C)$
Cos. A	a·b·c	$\dfrac{b^2 + c^2 - a^2}{2bc}$
Sin. C	c·A·a	$\dfrac{c \times \text{Sin. A}}{a}$
Cot. B	a·C·b	$\dfrac{a \times \text{csc C}}{b} - \text{Cot. C}$
c	b·C·B	$b \times \text{Sin. C} \times \text{csc B}$

0°

M	Sine	Cosine	Tan.	Cotan.	Secant	Cosec.	M
0	.00000	1.0000	.00000	Infinite	1.0000	Infinite	60
1	.00029	.0000	.00029	3437.7	.0000	3437.7	59
2	.00058	.0000	.00058	1718.9	.0000	1718.9	58
3	.00087	.0000	.00087	1145.9	.0000	1145.9	57
4	.00116	.0000	.00116	859.44	.0000	859.44	56
5	.00145	1.0000	.00145	687.55	1.0000	687.55	55
6	.00174	.0000	.00174	572.96	.0000	572.96	54
7	.00204	.0000	.00204	491.11	.0000	491.11	53
8	.00233	.0000	.00233	429.72	.0000	429.72	52
9	.00262	.0000	.00262	381.97	.0000	381.97	51
10	.00291	.99999	.00291	343.77	1.0000	343.77	50
11	.00320	.99999	.00320	312.52	.0000	312.52	49
12	.00349	.99999	.00349	286.48	.0000	286.48	48
13	.00378	.99999	.00378	264.44	.0000	264.44	47
14	.00407	.99999	.00407	245.55	.0000	245.55	46
15	.00436	.99999	.00436	229.18	1.0000	229.18	45
16	.00465	.99999	.00465	214.86	.0000	214.86	44
17	.00494	.99999	.00494	202.22	.0000	202.22	43
18	.00524	.99999	.00524	190.98	.0000	190.99	42
19	.00553	.99998	.00553	180.93	.0000	180.93	41
20	.00582	.99998	.00582	171.88	1.0000	171.89	40
21	.00611	.99998	.00611	163.70	.0000	163.70	39
22	.00640	.99998	.00640	156.26	.0000	156.26	38
23	.00669	.99998	.00669	149.46	.0000	149.47	37
24	.00698	.99997	.00698	143.24	.0000	143.24	36
25	.00727	.99997	.00727	137.51	1.0000	137.51	35
26	.00756	.99997	.00756	132.22	.0000	132.22	34
27	.00785	.99997	.00785	127.32	.0000	127.32	33
28	.00814	.99997	.00814	122.77	.0000	122.78	32
29	.00843	.99996	.00844	118.54	.0000	118.54	31
30	.00873	.99996	.00873	114.59	1.0000	114.59	30
31	.00902	.99996	.00902	110.89	.0000	110.90	29
32	.00931	.99996	.00931	107.43	.0000	107.43	28
33	.00960	.99995	.00960	104.17	.0000	104.17	27
34	.00989	.99995	.00989	101.11	.0000	101.11	26
35	.01018	.99995	.01018	98.218	1.0000	98.223	25
36	.01047	.99994	.01047	95.489	.0000	95.495	24
37	.01076	.99994	.01076	92.908	.0000	92.914	23
38	.01105	.99994	.01105	90.463	.0001	90.469	22
39	.01134	.99993	.01134	88.143	.0001	88.149	21
40	.01163	.99993	.01164	85.940	1.0001	85.946	20
41	.01193	.99993	.01193	83.843	.0001	83.849	19
42	.01222	.99992	.01222	81.847	.0001	81.853	18
43	.01251	.99992	.01251	79.943	.0001	79.950	17
44	.01280	.99992	.01280	78.126	.0001	78.133	16
45	.01309	.99991	.01309	76.390	1.0001	76.396	15
46	.01338	.99991	.01338	74.729	.0001	74.736	14
47	.01367	.99991	.01367	73.139	.0001	73.146	13
48	.01396	.99990	.01396	71.615	.0001	71.622	12
49	.01425	.99990	.01425	70.153	.0001	70.160	11
50	.01454	.99989	.01454	68.750	1.0001	68.757	10
51	.01483	.99989	.01484	67.402	.0001	67.409	9
52	.01512	.99988	.01513	66.105	.0001	66.113	8
53	.01542	.99988	.01542	64.858	.0001	64.866	7
54	.01571	.99988	.01571	63.657	.0001	63.664	6
55	.01600	.99987	.01600	62.499	1.0001	62.507	5
56	.01629	.99987	.01629	61.383	.0001	61.391	4
57	.01658	.99987	.01658	60.306	.0001	60.314	3
58	.01687	.99986	.01687	59.266	.0001	59.274	2
59	.01716	.99985	.01716	58.261	.0001	58.270	1
60	.01745	.99985	.01745	57.290	1.0001	57.299	0

M	Cosine	Sine	Cotan.	Tan.	Cosec.	Secant	M

89°

1°

M	Sine	Cosine	Tan.	Cotan.	Secant	Cosec.	M
0	.01745	.99985	.01745	57.290	1.0001	57.299	60
1	.01774	.99984	.01775	56.350	.0001	56.359	59
2	.01803	.99984	.01804	55.441	.0001	55.450	58
3	.01832	.99983	.01833	54.561	.0002	54.570	57
4	.01861	.99983	.01862	53.708	.0002	53.718	56
5	.01891	.99982	.01891	52.882	1.0002	52.891	55
6	.01920	.99981	.01920	52.081	.0002	52.090	54
7	.01949	.99981	.01949	51.303	.0002	51.313	53
8	.01978	.99980	.01978	50.548	.0002	50.558	52
9	.02007	.99980	.02007	49.816	.0002	49.826	51
10	.02036	.99979	.02036	49.104	1.0002	49.114	50
11	.02065	.99979	.02066	48.412	.0002	48.422	49
12	.02094	.99978	.02095	47.739	.0002	47.750	48
13	.02123	.99977	.02124	47.085	.0002	47.096	47
14	.02152	.99977	.02153	46.449	.0002	46.460	46
15	.02181	.99976	.02182	45.829	1.0002	45.840	45
16	.02210	.99975	.02211	45.226	.0002	45.237	44
17	.02240	.99975	.02240	44.638	.0002	44.650	43
18	.02269	.99974	.02269	44.066	.0002	44.077	42
19	.02298	.99974	.02298	43.508	.0003	43.520	41
20	.02326	.99973	.02327	42.964	1.0003	42.976	40
21	.02356	.99972	.02357	42.433	.0003	42.445	39
22	.02385	.99971	.02386	41.916	.0003	41.928	38
23	.02414	.99971	.02415	41.410	.0003	41.423	37
24	.02443	.99970	.02444	40.917	.0003	40.930	36
25	.02472	.99969	.02473	40.436	1.0003	40.448	35
26	.02501	.99969	.02502	39.965	.0003	39.978	34
27	.02530	.99968	.02531	39.506	.0003	39.518	33
28	.02559	.99967	.02560	39.057	.0003	39.069	32
29	.02589	.99966	.02589	38.618	.0003	38.631	31
30	.02618	.99966	.02618	38.188	1.0003	38.201	30
31	.02647	.99965	.02648	37.769	.0003	37.782	29
32	.02676	.99964	.02677	37.358	.0003	37.371	28
33	.02705	.99963	.02706	36.956	.0004	36.969	27
34	.02734	.99963	.02735	36.563	.0004	36.576	26
35	.02763	.99962	.02764	36.177	1.0004	36.191	25
36	.02792	.99961	.02793	35.800	.0004	35.814	24
37	.02821	.99960	.02822	35.431	.0004	35.445	23
38	.02850	.99959	.02851	35.069	.0004	35.084	22
39	.02879	.99958	.02880	34.715	.0004	34.729	21
40	.02908	.99958	.02910	34.368	1.0004	34.382	20
41	.02937	.99957	.02939	34.027	.0004	34.042	19
42	.02967	.99956	.02968	33.693	.0004	33.708	18
43	.02996	.99955	.02997	33.366	.0004	33.381	17
44	.03025	.99954	.03026	33.045	.0004	33.060	16
45	.03054	.99953	.03055	32.730	1.0005	32.745	15
46	.03083	.99952	.03084	32.421	.0005	32.437	14
47	.03112	.99951	.03113	32.118	.0005	32.134	13
48	.03141	.99951	.03143	31.820	.0005	31.836	12
49	.03170	.99950	.03172	31.528	.0005	31.544	11
50	.03199	.99949	.03201	31.241	1.0005	31.257	10
51	.03228	.99948	.03230	30.960	.0005	30.976	9
52	.03257	.99947	.03259	30.683	.0005	30.699	8
53	.03286	.99946	.03288	30.411	.0005	30.428	7
54	.03315	.99945	.03317	30.145	.0005	30.161	6
55	.03344	.99944	.03346	29.882	1.0005	29.899	5
56	.03374	.99943	.03375	29.624	.0006	29.641	4
57	.03403	.99942	.03405	29.371	.0006	29.388	3
58	.03432	.99941	.03434	29.122	.0006	29.139	2
59	.03461	.99940	.03463	28.877	.0006	28.894	1
60	.03490	.99939	.03492	28.636	1.0006	28.654	0

M	Cosine	Sine	Cotan.	Tan.	Cosec.	Secant	M

88°

2°

M	Sine	Cosine	Tan.	Cotan.	Secant	Cosec.	M
0	.03490	.99939	.03492	28.636	1.0006	28.654	60
1	.03519	.99938	.03521	28.399	.0006	28.417	59
2	.03548	.99937	.03550	28.166	.0006	28.184	58
3	.03577	.99936	.03579	27.937	.0006	27.955	57
4	.03606	.99935	.03608	27.712	.0006	27.730	56
5	.03635	.99934	.03638	27.490	1.0007	27.508	55
6	.03664	.99933	.03667	27.271	.0007	27.290	54
7	.03693	.99932	.03696	27.056	.0007	27.075	53
8	.03722	.99931	.03725	26.845	.0007	26.864	52
9	.03751	.99930	.03754	26.637	.0007	26.655	51
10	.03781	.99928	.03783	26.432	1.0007	26.450	50
11	.03810	.99927	.03812	26.230	.0007	26.249	49
12	.03839	.99926	.03842	26.031	.0007	26.050	48
13	.03868	.99925	.03871	25.835	.0007	25.854	47
14	.03897	.99924	.03900	25.642	.0008	25.661	46
15	.03926	.99923	.03929	25.452	1.0008	25.471	45
16	.03955	.99922	.03958	25.264	.0008	25.284	44
17	.03984	.99921	.03987	25.080	.0008	25.100	43
18	.04013	.99919	.04016	24.898	.0008	24.918	42
19	.04042	.99918	.04045	24.718	.0008	24.739	41
20	.04071	.99917	.04075	24.542	1.0008	24.562	40
21	.04100	.99916	.04104	24.367	.0008	24.388	39
22	.04129	.99915	.04133	24.196	.0008	24.216	38
23	.04158	.99913	.04162	24.026	.0009	24.047	37
24	.04187	.99912	.04191	23.859	.0009	23.880	36
25	.04217	.99911	.04220	23.694	1.0009	23.716	35
26	.04246	.99910	.04249	23.532	.0009	23.553	34
27	.04275	.99908	.04279	23.372	.0009	23.393	33
28	.04304	.99907	.04308	23.214	.0009	23.235	32
29	.04333	.99906	.04337	23.058	.0009	23.079	31
30	.04362	.99905	.04366	22.904	1.0009	22.925	30
31	.04391	.99903	.04395	22.752	.0010	22.774	29
32	.04420	.99902	.04424	22.602	.0010	22.624	28
33	.04449	.99901	.04453	22.454	.0010	22.476	27
34	.04478	.99900	.04483	22.308	.0010	22.330	26
35	.04507	.99898	.04512	22.164	1.0010	22.186	25
36	.04536	.99897	.04541	22.022	.0010	22.044	24
37	.04565	.99896	.04570	21.881	.0010	21.904	23
38	.04594	.99894	.04599	21.742	.0010	21.765	22
39	.04623	.99893	.04628	21.606	.0011	21.629	21
40	.04652	.99892	.04657	21.470	1.0011	21.494	20
41	.04681	.99890	.04687	21.337	.0011	21.360	19
42	.04711	.99889	.04716	21.205	.0011	21.228	18
43	.04740	.99888	.04745	21.075	.0011	21.098	17
44	.04769	.99886	.04774	20.946	.0011	20.970	16
45	.04798	.99885	.04803	20.819	1.0011	20.843	15
46	.04827	.99883	.04832	20.693	.0012	20.717	14
47	.04856	.99882	.04862	20.569	.0012	20.593	13
48	.04885	.99881	.04891	20.446	.0012	20.471	12
49	.04914	.99879	.04920	20.325	.0012	20.350	11
50	.04943	.99878	.04949	20.205	1.0012	20.230	10
51	.04972	.99876	.04978	20.087	.0012	20.112	9
52	.05001	.99875	.05007	19.970	.0012	19.995	8
53	.05030	.99873	.05037	19.854	.0013	19.880	7
54	.05059	.99872	.05066	19.740	.0013	19.766	6
55	.05088	.99870	.05095	19.627	1.0013	19.653	5
56	.05117	.99869	.05124	19.515	.0013	19.541	4
57	.05146	.99867	.05153	19.405	.0013	19.431	3
58	.05175	.99866	.05182	19.236	.0013	19.322	2
59	.05204	.99864	.05212	19.188	.0013	19.214	1
60	.05234	.99863	.05241	19.081	1.0014	19.107	0
M	Cosine	Sine	Cotan.	Tan.	Cosec.	Secant	M

87°

31

3°

M	Sine	Cosine	Tan.	Cotan.	Secant	Cosec.	M
0	.05234	.99863	.05241	19.081	1.0014	19.107	60
1	.05263	.99861	.05270	18.975	.0014	19.002	59
2	.05292	.99860	.05299	18.871	.0014	18.897	58
3	.05321	.99858	.05328	18.768	.0014	18.794	57
4	.05350	.99857	.05357	18.665	.0014	18.692	56
5	.05379	.99855	.05387	18.564	1.0014	18.591	55
6	.05408	.99854	.05416	18.464	.0015	18.491	54
7	.05437	.99852	.05445	18.365	.0015	18.393	53
8	.05466	.99850	.05474	18.268	.0015	18.295	52
9	.05495	.99849	.05503	18.171	.0015	18.198	51
10	.05524	.99847	.05532	18.075	1.0015	18.103	50
11	.05553	.99846	.05562	17.980	.0015	18.008	49
12	.05582	.99844	.05591	17.886	.0016	17.914	48
13	.05611	.99842	.05620	17.793	.0016	17.821	47
14	.05640	.99841	.05649	17.701	.0016	17.730	46
15	.05669	.99839	.05678	17.610	1.0016	17.639	45
16	.05698	.99837	.05707	17.520	.0016	17.549	44
17	.05727	.99836	.05737	17.431	.0016	17.460	43
18	.05756	.99834	.05766	17.343	.0017	17.372	42
19	.05785	.99832	.05795	17.256	.0017	17.285	41
20	.05814	.99831	.05824	17.169	1.0017	17.198	40
21	.05843	.99829	.05853	17.084	.0017	17.113	39
22	.05872	.99827	.05883	16.999	.0017	17.028	38
23	.05902	.99826	.05912	16.915	.0017	16.944	37
24	.05931	.99824	.05941	16.832	.0018	16.861	36
25	.05960	.99822	.05970	16.750	1.0018	16.779	35
26	.05989	.99820	.05999	16.668	.0018	16.698	34
27	.06018	.99819	.06029	16.587	.0018	16.617	33
28	.06047	.99817	.06058	16.507	.0018	16.538	32
29	.06076	.99815	.06087	16.428	.0018	16.459	31
30	.06105	.99813	.06116	16.350	1.0019	16.380	30
31	.06134	.99812	.06145	16.272	.0019	16.303	29
32	.06163	.99810	.06175	16.195	.0019	16.226	28
33	.06192	.99808	.06204	16.119	.0019	16.150	27
34	.06221	.99806	.06233	16.043	.0019	16.075	26
35	.06250	.99804	.06262	15.969	1.0019	16.000	25
36	.06279	.99803	.06291	15.894	.0020	15.926	24
37	.06308	.99801	.06321	15.821	.0020	15.853	23
38	.06337	.99799	.06350	15.748	.0020	15.780	22
39	.06366	.99797	.06379	15.676	.0020	15.708	21
40	.06395	.99795	.06408	15.605	1.0020	15.637	20
41	.06424	.99793	.06437	15.534	.0021	15.566	19
42	.06453	.99791	.06467	15.464	.0021	15.496	18
43	.06482	.99790	.06496	15.394	.0021	15.427	17
44	.06511	.99788	.06525	15.325	.0021	15.358	16
45	.06540	.99786	.06554	15.257	1.0021	15.290	15
46	.06569	.99784	.06583	15.189	.0022	15.222	14
47	.06598	.99782	.06613	15.122	.0022	15.155	13
48	.06627	.99780	.06642	15.056	.0022	15.089	12
49	.06656	.99778	.06671	14.990	.0022	15.023	11
50	.06685	.99776	.06700	14.924	1.0022	14.958	10
51	.06714	.99774	.06730	14.860	.0023	14.893	9
52	.06743	.99772	.06759	14.795	.0023	14.829	8
53	.06772	.99770	.06788	14.732	.0023	14.765	7
54	.06801	.99768	.06817	14.668	.0023	14.702	6
55	.06830	.99766	.06846	14.606	1.0023	14.640	5
56	.06859	.99764	.06876	14.544	.0024	14.578	4
57	.06888	.99762	.06905	14.482	.0024	14.517	3
58	.06918	.99760	.06934	14.421	.0024	14.456	2
59	.06947	.99758	.06963	14.361	.0024	14.395	1
60	.06976	.99756	.06993	14.301	1.0024	14.335	0

M	Cosine	Sine	Cotan.	Tan.	Cosec.	Secant	M

86°

4°

M	Sine	Cosine	Tan.	Cotan.	Secant	Cosec.	M
0	.06976	.99756	.06993	14.301	1.0024	14.335	60
1	.07005	.99754	.07022	14.241	.0025	14.276	59
2	.07034	.99752	.07051	14.182	.0025	14.217	58
3	.07063	.99750	.07080	14.123	.0025	14.159	57
4	.07092	.99748	.07110	14.065	.0025	14.101	56
5	.07121	.99746	.07139	14.008	1.0025	14.043	55
6	.07150	.99744	.07168	13.951	.0026	13.986	54
7	.07179	.99742	.07197	13.894	.0026	13.930	53
8	.07208	.99740	.07226	13.838	.0026	13.874	52
9	.07237	.99738	.07256	13.782	.0026	13.818	51
10	.07266	.99736	.07285	13.727	1.0026	13.763	50
11	.07295	.99733	.07314	13.672	.0027	13.708	49
12	.07324	.99731	.07343	13.617	.0027	13.654	48
13	.07353	.99729	.07373	13.533	.0027	13.600	47
14	.07382	.99727	.07402	13.510	.0027	13.547	46
15	.07411	.99725	.07431	13.457	1.0027	13.494	45
16	.07440	.99723	.07460	13.404	.0028	13.441	44
17	.07469	.99721	.07490	13.351	.0028	13.389	43
18	.07498	.99718	.07519	13.299	.0028	13.337	42
19	.07527	.99716	.07548	13.248	.0028	13.286	41
20	.07556	.99714	.07577	13.197	1.0029	13.235	40
21	.07585	.99712	.07607	13.146	.0029	13.184	39
22	.07614	.99710	.07636	13.096	.0029	13.134	38
23	.07643	.99707	.07665	13.046	.0029	13.084	37
24	.07672	.99705	.07694	12.996	.0029	13.034	36
25	.07701	.99703	.07724	12.947	1.0030	12.985	35
26	.07730	.99701	.07753	12.898	.0030	12.937	34
27	.07759	.99698	.07782	12.849	.0030	12.888	33
28	.07788	.99696	.07812	12.801	.0030	12.840	32
29	.07817	.99694	.07841	12.754	.0031	12.793	31
30	.07846	.99692	.07870	12.706	1.0031	12.745	30
31	.07875	.99689	.07899	12.659	.0031	12.698	29
32	.07904	.99687	.07929	12.612	.0031	12.652	28
33	.07933	.99685	.07958	12.566	.0032	12.606	27
34	.07962	.99682	.07987	12.520	.0032	12.560	26
35	.07991	.99680	.08016	12.474	1.0032	12.514	25
36	.08020	.99678	.08046	12.429	.0032	12.469	24
37	.08049	.99675	.08075	12.384	.0032	12.424	23
38	.08078	.99673	.08104	12.339	.0033	12.379	22
39	.08107	.99671	.08134	12.295	.0033	12.335	21
40	.08136	.99668	.08163	12.250	1.0033	12.291	20
41	.08165	.99666	.08192	12.207	.0033	12.248	19
42	.08194	.99664	.08221	12.163	.0034	12.204	18
43	.08223	.99661	.08251	12.120	.0034	12.161	17
44	.08252	.99659	.08280	12.077	.0034	12.118	16
45	.08281	.99656	.08309	12.035	1.0·34	12.076	15
46	.08310	.99654	.08339	11.992	.0035	12.034	14
47	.08339	.99652	.08368	11.950	.0035	11.992	13
48	.08368	.99649	.08397	11.909	.0035	11.950	12
49	.08397	.99647	.08426	11.867	.0035	11.909	11
50	.08426	.99644	.08456	11.826	1.0036	11.868	10
51	.08455	.99642	.08485	11.785	.0036	11.828	9
52	.08484	.99639	.08514	11.745	.0036	11.787	8
53	.08513	.99637	.08544	11.704	.0036	11.747	7
54	.08542	.99634	.08573	11.664	.0037	11.707	6
55	.08571	.99632	.08602	11.625	1.0037	11.668	5
56	.08600	.99629	.08632	11.585	.0037	11.628	4
57	.08629	.99627	.08661	11.546	.0037	11.589	3
58	.08658	.99624	.08690	11.507	.0038	11.550	2
59	.08687	.99622	.08719	11.468	.0038	11.512	1
60	.08715	.99619	.08749	11.430	1.0038	11.474	0
M	Cosine	Sine	Cotan.	Tan.	Cosec.	Secant	M

85°

5°

M	Sine	Cosine	Tan	Cotan.	Secant	Cosec.	M
0	.08715	.99619	.08749	11.430	1.0038	11.474	60
1	.08744	.99617	.08778	11.392	.0038	11.436	59
2	.08773	.99614	.08807	11.354	.0039	11.398	58
3	.08802	.99612	.08837	11.316	.0039	11.360	57
4	.08831	.99609	.08866	11.279	.0039	11.323	56
5	.08860	.99607	.08895	11.242	1.0039	11.286	55
6	.08889	.99604	.08925	11.205	.0040	11.249	54
7	.08918	.99601	.08954	11.168	.0040	11.213	53
8	.08947	.99599	.08988	11.132	.0040	11.176	52
9	.08976	.99596	.09013	11.095	.0040	11.140	51
10	.09005	.99594	.09042	11.059	1.0041	11.104	50
11	.09034	.99591	.09071	11.024	.0041	11.069	49
12	.09063	.99588	.09101	10.988	.0041	11.033	48
13	.09092	.99586	.09130	10.953	.0041	10.998	47
14	.09121	.99583	.09159	10.918	.0042	10.963	46
15	.09150	.99580	.09189	10.883	1.0042	10.929	45
16	.09179	.99578	.09218	10.848	.0042	10.894	44
17	.09208	.99575	.09247	10.814	.0043	10.860	43
18	.09237	.99572	.09277	10.780	.0043	10.826	42
19	.09266	.99570	.09306	10.746	.0043	10.792	41
20	.09295	.99567	.09335	10.712	1.0043	10.758	40
21	.09324	.99564	.09365	10.678	.0044	10.725	39
22	.09353	.99562	.09394	10.645	.0044	10.692	38
23	.09382	.99559	.09423	10.612	.0044	10.659	37
24	.09411	.99556	.09453	10.579	.0044	10.626	36
25	.09440	.99553	.09482	10.546	1.0045	10.593	35
26	.09469	.99551	.09511	10.514	.0045	10.561	34
27	.09498	.99548	.09541	10.481	.0045	10.529	33
28	.09527	.99545	.09570	10.449	.0046	10.497	32
29	.09556	.99542	.09599	10.417	.0046	10.465	31
30	.09584	.99540	.09629	10.385	1.0046	10.433	30
31	.09613	.99537	.09658	10.354	.0046	10.402	29
32	.09642	.99534	.09688	10.322	.0047	10.371	28
33	.09671	.99531	.09717	10.291	.0047	10.340	27
34	.09700	.99528	.09746	10.260	.0047	10.309	26
35	.09729	.99525	.09776	10.229	1.0048	10.278	25
36	.09758	.99523	.09805	10.199	.0048	10.248	24
37	.09787	.99520	.09834	10.168	.0048	10.217	23
38	.09816	.99517	.09864	10.138	.0048	10.187	22
39	.09845	.99514	.09893	10.108	.0049	10.157	21
40	.09874	.99511	.09922	10.078	1.0049	10.127	20
41	.09903	.99508	.09952	10.048	.0049	10.098	19
42	.09932	.99505	.09981	10.019	.0050	10.068	18
43	.09961	.99503	.10011	9.9893	.0050	10.039	17
44	.09990	.99500	.10040	9.9601	.0050	10.010	16
45	.10019	.99497	.10069	9.9310	1.0050	9.9812	15
46	.10048	.99494	.10099	9.9021	.0051	9.9525	14
47	.10077	.99491	.10128	9.8734	.0051	9.9239	13
48	.10106	.99488	.10158	9.8448	.0051	9.8955	12
49	.10134	.99485	.10187	9.8164	.0052	9.8672	11
50	.10163	.99482	.10216	9.7882	1.0052	9.8391	10
51	.10192	.99479	.10246	9.7601	.0052	9.8112	9
52	.10221	.99476	.10275	9.7322	.0053	9.7834	8
53	.10250	.99473	.10305	9.7044	.0053	9.7558	7
54	.10279	.99470	.10334	9.6768	.0053	9.7283	6
55	.10308	.99467	.10363	9.6493	1.0053	9.7010	5
56	.10337	.99464	.10393	9.6220	.0054	9.6739	4
57	.10366	.99461	.10422	9.5949	.0054	9.6469	3
58	.10395	.99458	.10452	9.5679	.0054	9.6200	2
59	.10424	.99455	.10481	9.5411	.0055	9.5933	1
60	.10453	.99452	.10510	9.5144	1.0055	9.5668	0

| M | Cosine | Sine | Cotan. | Tan. | Cosec. | Secant | M |

84°

34

6°

M	Sine	Cosine	Tan.	Cotsn.	Secant	Cosec.	M
0	.10453	.99452	.10510	9.5144	1.0055	9.5668	60
1	.10482	.99449	.10540	.4878	.0055	.5404	59
2	.10511	.99446	.10569	.4614	.0056	.5141	58
3	.10540	.99443	.10599	.4351	.0056	.4880	57
4	.10568	.99440	.10628	.4090	.0056	.4620	56
5	.10597	.99437	.10657	9.3831	1.0057	9.4362	55
6	.10626	.99434	.10687	.3572	.0057	.4105	54
7	.10655	.99431	.10716	.3315	.0057	.3850	53
8	.10684	.99428	.10746	.3060	.0057	.3596	52
9	.10713	.99424	.10775	.2806	.0058	.3343	51
10	.10742	.99421	.10805	9.2553	1.0058	9.3092	50
11	.10771	.99418	.10834	.2302	.0058	.2842	49
12	.10800	.99415	.10863	.2051	.0059	.2593	48
13	.10829	.99412	.10893	.1803	.0059	.2346	47
14	.10858	.99409	.10922	.1555	.0059	.2100	46
15	.10887	.99406	.10952	9.1309	1.0060	9.1855	45
16	.10916	.99402	.10981	.1064	.0060	.1612	44
17	.10944	.99399	.11011	.0821	.0060	.1370	43
18	.10973	.99396	.11040	.0579	.0061	.1129	42
19	.11002	.99393	.11069	.0338	.0061	.0890	41
20	.11031	.99390	.11099	9.0098	1.0061	9.0651	40
21	.11060	.99386	.11128	8.9860	.0062	.0414	39
22	.11089	.99383	.11158	.9623	.0062	.0179	38
23	.11118	.99380	.11187	.9387	.0062	8.9944	37
24	.11147	.99377	.11217	.9152	.0063	.9711	36
25	.11176	.99373	.11246	8.8918	1.0063	8.9479	35
26	.11205	.99370	.11276	.8686	.0063	.9248	34
27	.11234	.99367	.11305	.8455	.0064	.9018	33
28	.11262	.99364	.11335	.8225	.0064	.8790	32
29	.11291	.99360	.11364	.7996	.0064	.8563	31
30	.11320	.99357	.11393	8.7769	1.0065	8.8337	30
31	.11349	.99354	.11423	.7542	.0065	.8112	29
32	.11378	.99350	.11452	.7317	.0065	.7888	28
33	.11407	.99347	.11482	.7093	.0066	.7665	27
34	.11436	.99344	.11511	.6870	.0066	.7444	26
35	.11465	.99341	.11541	8.6648	1.0066	8.7223	25
36	.11494	.99337	.11570	.6427	.0067	.7004	24
37	.11523	.99334	.11600	.6208	.0067	.6786	23
38	.11551	.99330	.11629	.5989	.0067	.6569	22
39	.11580	.99327	.11659	.5772	.0068	.6353	21
40	.11609	.99324	.11688	8.5555	1.0068	8.6138	20
41	.11638	.99320	.11718	.5340	.0068	.5924	19
42	.11667	.99317	.11747	.5126	.0069	.5711	18
43	.11696	.99314	.11777	.4913	.0069	.5499	17
44	.11725	.99310	.11806	.4701	.0069	.5289	16
45	.11754	.99307	.11836	8.4489	1.0070	8.5079	15
46	.11783	.99303	.11865	.4279	.0070	.4871	14
47	.11811	.99300	.11895	.4070	.0070	.4663	13
48	.11840	.99296	.11924	.3862	.0071	.4457	12
49	.11869	.99293	.11954	.3655	.0071	.4251	11
50	.11898	.99290	.11983	8.3449	1.0071	8.4046	10
51	.11927	.99286	.12013	.3244	.0072	.3843	9
52	.11956	.99283	.12042	.3040	.0072	.3640	8
53	.11985	.99279	.12072	.2837	.0073	.3439	7
54	.12014	.99276	.12101	.2635	.0073	.3238	6
55	.12042	.99272	.12131	8.2434	1.0073	8.3039	5
56	.12071	.99269	.12160	.2234	.0074	.2840	4
57	.12100	.99265	.12190	.2035	.0074	.2642	3
58	.12129	.99262	.12219	.1837	.0074	.2446	2
59	.12158	.99258	.12249	.1640	.0075	.2250	1
60	.12187	.99255	.12278	8.1443	1.0075	8.2055	0

M	Cosine	Sine	Cotan.	Tan.	Cosec.	Secant	M

83°

7°

M	Sine	Cosine	Tan.	Cotan.	Secant	Cosec.	M
0	.12187	.99255	.12278	8.1443	1.0075	8.2055	60
1	.12216	.99251	.12308	.1248	.0075	.1861	59
2	.12245	.99247	.12337	.1053	.0076	.1668	58
3	.12273	.99244	.12367	.0860	.0076	.1476	57
4	.12302	.99240	.12396	.0667	.0076	.1285	56
5	.12331	.99237	.12426	8.0476	1.0077	8.1094	55
6	.12360	.99233	.12456	.0285	.0077	.0905	54
7	.12389	.99229	.12485	.0095	.0078	.0717	53
8	.12418	.99226	.12515	7.9906	.0078	.0529	52
9	.12447	.99222	.12544	.9717	.0078	.0342	51
10	.12476	.99219	.12574	7.9530	1.0079	8.0156	50
11	.12504	.99215	.12603	.9344	.0079	7.9971	49
12	.12533	.99211	.12633	.9158	.0079	.9787	48
13	.12562	.99208	.12662	.8973	.0080	.9604	47
14	.12591	.99204	.12692	.8789	.0080	.9421	46
15	.12620	.99200	.12722	7.8606	1.0080	7.9240	45
16	.12649	.99197	.12751	.8424	.0081	.9059	44
17	.12678	.99193	.12781	.8243	.0081	.8879	43
18	.12706	.99189	.12810	.8062	.0082	.8700	42
19	.12735	.99186	.12840	.7882	.0082	.8522	41
20	.12764	.99182	.12869	7.7703	1.0082	7.8344	40
21	.12793	.99178	.12899	.7525	.0083	.8168	39
22	.12822	.99174	.12928	.7348	.0083	.7992	38
23	.12851	.99171	.12958	.7171	.0084	.7817	37
24	.12879	.99167	.12988	.6996	.0084	.7642	36
25	.12908	.99163	.13017	7.6821	1.0084	7.7469	35
26	.12937	.99160	.13047	.6646	.0085	.7296	34
27	.12966	.99156	.13076	.6473	.0085	.7124	33
28	.12995	.99152	.13106	.6300	.0085	.6953	32
29	.13024	.99148	.13136	.6129	.0086	.6783	31
30	.13053	.99144	.13165	7.5957	1.0086	7.6613	30
31	.13081	.99141	.13195	.5787	.0087	.6444	29
32	.13110	.99137	.13224	.5617	.0087	.6276	28
33	.13139	.99133	.13254	.5449	.0087	.6108	27
34	.13168	.99129	.13284	.5280	.0088	.5942	26
35	.13197	.99125	.13313	7.5113	1.0088	7.5776	25
36	.13226	.99121	.13343	.4946	.0089	.5611	24
37	.13254	.99118	.13372	.4780	.0089	.5446	23
38	.13283	.99114	.13402	.4615	.0089	.5282	22
39	.13312	.99110	.13432	.4451	.0090	.5119	21
40	.13341	.99106	.13461	7.4287	1.0090	7.4957	20
41	.13370	.99102	.13491	.4124	.0090	.4795	19
42	.13399	.99098	.13520	.3961	.0091	.4634	18
43	.13427	.99094	.13550	.3800	.0091	.4474	17
44	.13456	.99090	.13580	.3639	.0092	.4315	16
45	.13485	.99086	.13609	7.3479	1.0092	7.4156	15
46	.13514	.99083	.13639	.3319	.0092	.3998	14
47	.13543	.99079	.13669	.3160	.0093	.3840	13
48	.13571	.99075	.13698	.3002	.0093	.3683	12
49	.13600	.99071	.13728	.2844	.0094	.3527	11
50	.13629	.99067	.13757	7.2687	1.0094	7.3372	10
51	.13658	.99063	.13787	.2531	.0094	.3217	9
52	.13687	.99059	.13817	.2375	.0095	.3063	8
53	.13716	.99055	.13846	.2220	.0095	.2909	7
54	.13744	.99051	.13876	.2066	.0096	.2757	6
55	.13773	.99047	.13906	7.1912	1.0096	7.2604	5
56	.13802	.99043	.13935	.1759	.0097	.2453	4
57	.13831	.99039	.13965	.1607	.0097	.2302	3
58	.13860	.99035	.13995	.1455	.0097	.2152	2
59	.13888	.99031	.14024	.1304	.0098	.2002	1
60	.13917	.99027	.14054	7.1154	1.0098	7.1853	0
M	Cosine	Sine	Cotan.	Tan.	Cosec.	Secant	M

82°

36

8°

M	Sine	Cosine	Tan.	Cotan.	Secant	Cosec.	M
0	.13917	.99027	.14054	7.1154	1.0098	7.1853	60
1	.13946	.99023	.14084	.1004	.0099	.1704	59
2	.13975	.99019	.14113	.0854	.0099	.1557	58
3	.14004	.99015	.14143	.0706	.0099	.1409	57
4	.14032	.99010	.14173	.0558	.0100	.1263	56
5	.14061	.99006	.14202	7.0410	1.0100	7.1117	55
6	.14090	.99002	.14232	.0264	.0101	.0972	54
7	.14119	.98998	.14262	.0117	.0101	.0827	53
8	.14148	.98994	.14291	6.9972	.0102	.0683	52
9	.14176	.98990	.14321	.9827	.0102	.0539	51
10	.14205	.98986	.14351	6.9682	1.0102	7.0396	50
11	.14234	.98982	.14380	.9538	.0103	.0254	49
12	.14263	.98978	.14410	.9395	.0103	.0112	48
13	.14292	.98973	.14440	.9252	.0104	6.9971	47
14	.14320	.98969	.14470	.9110	.0104	.9830	46
15	.14349	.98965	.14499	6.8969	1.0104	6.9690	45
16	.14378	.98961	.14529	.8828	.0105	.9550	44
17	.14407	.98957	.14559	.8687	.0105	.9411	43
18	.14436	.98952	.14588	.8547	.0106	.9273	42
19	.14464	.98948	.14618	.8408	.0106	.9135	41
20	.14493	.98944	.14648	6.8269	1.0107	6.8998	40
21	.14522	.98940	.14677	.8131	.0107	.8861	39
22	.14551	.98936	.14707	.7993	.0107	.8725	38
23	.14579	.98931	.14737	.7856	.0108	.8589	37
24	.14608	.98927	.14767	.7720	.0108	.8454	36
25	.14637	.98923	.14796	6.7584	1.0109	6.8320	35
26	.14666	.98919	.14826	.7448	.0109	.8185	34
27	.14695	.98914	.14856	.7313	.0110	.8052	33
28	.14723	.98910	.14886	.7179	.0110	.7919	32
29	.14752	.98906	.14915	.7045	.0111	.7787	31
30	.14781	.98901	.14945	6.6911	1.0111	6.7655	30
31	.14810	.98897	.14975	.6779	.0111	.7523	29
32	.14838	.98893	.15004	.6646	.0112	.7392	28
33	.14867	.98889	.15034	.6514	.0112	.7262	27
34	.14896	.98884	.15064	.6383	.0113	.7132	26
35	.14925	.98880	.15094	6.6252	1.0113	6.7003	25
36	.14953	.98876	.15123	.6122	.0114	.6874	24
37	.14982	.98871	.15153	.5992	.0114	.6745	23
38	.15011	.98867	.15183	.5863	.0115	.6617	22
39	.15040	.98862	.15213	.5734	.0115	.6490	21
40	.15068	.98858	.15243	6.5605	1.0115	6.6363	20
41	.15097	.98854	.15272	.5478	.0116	.6237	19
42	.15126	.98849	.15302	.5350	.0116	.6111	18
43	.15155	.98845	.15332	.5223	.0117	.5985	17
44	.15183	.98840	.15362	.5097	.0117	.5860	16
45	.15212	.98836	.15391	6.4971	1.0118	6.5736	15
46	.15241	.98832	.15421	.4845	.0118	.5612	14
47	.15270	.98827	.15451	.4720	.0119	.5488	13
48	.15298	.98823	.15481	.4596	.0119	.5365	12
49	.15328	.98818	.15511	.4472	.0119	.5243	11
50	.15356	.98814	.15540	6.4348	1.0120	6.5121	10
51	.15385	.98809	.15570	.4225	.0120	.4999	9
52	.15413	.98805	.15600	.4103	.0121	.4878	8
53	.15442	.98800	.15630	.3980	.0121	.4757	7
54	.15471	.98796	.15659	.3859	.0122	.4637	6
55	.15500	.98791	.15689	6.3737	1.0122	6.4517	5
56	.15528	.98787	.15719	.3616	.0123	.4398	4
57	.15557	.98782	.15749	.3496	.0123	.4279	3
58	.15586	.98778	.15779	.3376	.0124	.4160	2
59	.15615	.98773	.15809	.3257	.0124	.4042	1
60	.15643	.98769	.15838	6.3137	1.0125	6.3924	0

M	Cosine	Sine	Cotan.	Tan.	Cosec.	Secant	M

81°

9°

M	Sine	Cosine	Tan.	Cotan.	Secant	Cosec.	M
0	.15643	.98769	.15838	6.3137	1.0125	6.3924	60
1	.15672	.98764	.15868	.3019	.0125	.3807	59
2	.15701	.98760	.15898	.2901	.0125	.3690	58
3	.15730	.98755	.15928	.2783	.0126	.3574	57
4	.15758	.98750	.15958	.2665	.0126	.3458	56
5	.15787	.98746	.15987	6.2548	1.0127	6.3343	55
6	.15816	.98741	.16017	.2432	.0127	.3228	54
7	.15844	.98737	.16047	.2316	.0128	.3113	53
8	.15873	.98732	.16077	.2200	.0128	.2999	52
9	.15902	.98727	.16107	.2085	.0129	.2885	51
10	.15931	.98723	.16137	6.1970	1.0129	6.2772	50
11	.15959	.98718	.16167	.1856	.0130	.2659	49
12	.15988	.98714	.16196	.1742	.0130	.2546	48
13	.16017	.98709	.16226	.1628	.0131	.2434	47
14	.16045	.98704	.16256	.1515	.0131	.2322	46
15	.16074	.98700	.16286	6.1402	1.0132	6.2211	45
16	.16103	.98695	.16316	.1290	.0132	.2100	44
17	.16132	.98690	.16346	.1178	.0133	.1990	43
18	.16160	.98685	.16376	.1066	.0133	.1880	42
19	.16189	.98681	.16405	.0955	.0134	.1770	41
20	.16218	.98676	.16435	6.0844	1.0134	6.1661	40
21	.16246	.98671	.16465	.0734	.0135	.1552	39
22	.16275	.98667	.16495	.0624	.0135	.1443	38
23	.16304	.98662	.16525	.0514	.0136	.1335	37
24	.16333	.98657	.16555	.0405	.0136	.1227	36
25	.16361	.98652	.16585	6.0296	1.0136	6.1120	35
26	.16390	.98648	.16615	.0188	.0137	.1013	34
27	.16419	.98643	.16644	.0080	.0137	.0906	33
28	.16447	.98638	.16674	5.9972	.0138	.0800	32
29	.16476	.98633	.16704	.9865	.0138	.0694	31
30	.16505	.98628	.16734	5.9758	1.0139	6.0588	30
31	.16533	.98624	.16764	.9651	.0139	.0483	29
32	.16562	.98619	.16794	.9545	.0140	.0379	28
33	.16591	.98614	.16824	.9439	.0140	.0274	27
34	.16619	.98609	.16854	.9333	.0141	.0170	26
35	.16648	.98604	.16884	5.9228	1.0141	6.0066	25
36	.16677	.98600	.16914	.9123	.0142	5.9963	24
37	.16705	.98595	.16944	.9019	.0142	.9860	23
38	.16734	.98590	.16973	.8915	.0143	.9758	22
39	.16763	.98585	.17003	.8811	.0143	.9655	21
40	.16791	.98580	.17033	5.8708	1.0144	5.9554	20
41	.16820	.98575	.17063	.8605	.0144	.9452	19
42	.16849	.98570	.17093	.8502	.0145	.9351	18
43	.16878	.98565	.17123	.8400	.0145	.9250	17
44	.16906	.98560	.17153	.8298	.0146	.9150	16
45	.16935	.98556	.17183	5.8196	1.0146	5.9049	15
46	.16964	.98551	.17213	.8095	.0147	.8950	14
47	.16992	.98546	.17243	.7994	.0147	.8850	13
48	.17021	.98541	.17273	.7894	.0148	.8751	12
49	.17050	.98536	.17303	.7794	.0148	.8652	11
50	.17078	.98531	.17333	5.7694	1.0149	5.8554	10
51	.17107	.98526	.17363	.7594	.0150	.8456	9
52	.17136	.98521	.17393	.7495	.0150	.8358	8
53	.17164	.98516	.17423	.7396	.0151	.8261	7
54	.17193	.98511	.17453	.7297	.0151	.8163	6
55	.17221	.98506	.17483	5.7199	1.0152	5.8067	5
56	.17250	.98501	.17513	.7101	.0152	.7970	4
57	.17279	.98496	.17543	.7004	.0153	.7874	3
58	.17307	.98491	.17573	.6906	.0153	.7778	2
59	.17336	.98486	.17603	.6809	.0154	.7683	1
60	.17365	.98481	.17633	5.6713	1.0154	5.7588	0

M	Cosine	Sine	Cotan.	Tan.	Cosec.	Secant	M

80°

38

10°

M	Sine	Cosine	Tan.	Cotn.	Secant	Cosec.	M
0	.17365	.98481	.17633	5.6713	1.0154	5.7588	60
1	.17393	.98476	.17663	.6616	.0155	.7493	59
2	.17422	.98471	.17693	.6520	.0155	.7398	58
3	.17451	.98465	.17723	.6425	.0156	.7304	57
4	.17479	.98460	.17753	.6329	.0156	.7210	56
5	.17508	.98455	.17783	5.6234	1.0157	5.7117	55
6	.17537	.98450	.17813	.6140	.0157	.7023	54
7	.17565	.98445	.17843	.6045	.0158	.6930	53
8	.17594	.98440	.17873	.5951	.0158	.6838	52
9	.17622	.98435	.17903	.5857	.0159	.6745	51
10	.17651	.98430	.17933	5.5764	1.0159	5.6653	50
11	.17680	.98425	.17963	.5670	.0160	.6561	49
12	.17708	.98419	.17993	.5578	.0160	.6470	48
13	.17737	.98414	.18023	.5485	.0161	.6379	47
14	.17766	.98409	.18053	.5393	.0162	.6288	46
15	.17794	.98404	.18083	5.5301	1.0162	5.6197	45
16	.17823	.98399	.18113	.5209	.0163	.6107	44
17	.17852	.98394	.18143	.5117	.0163	.6017	43
18	.17880	.98388	.18173	.5026	.0164	.5928	42
19	.17909	.98383	.18203	.4936	.0164	.5838	41
20	.17937	.98378	.18233	5.4845	1.0165	5.5749	40
21	.17966	.98373	.18263	.4755	.0165	.5660	39
22	.17995	.98368	.18293	.4665	.0166	.5572	38
23	.18023	.98362	.18323	.4575	.0166	.5484	37
24	.18052	.98357	.18353	.4486	.0167	.5396	36
25	.18080	.98352	.18383	5.4396	1.0167	5.5308	35
26	.18109	.98347	.18413	.4308	.0168	.5221	34
27	.18138	.98341	.18444	.4219	.0169	.5134	33
28	.18166	.98336	.18474	.4131	.0169	.5047	32
29	.18195	.98331	.18504	.4043	.0170	.4960	31
30	.18223	.98325	.18534	5.3955	1.0170	5.4874	30
31	.18252	.98320	.18564	.3868	.0171	.4788	29
32	.18281	.98315	.18594	.3780	.0171	.4702	28
33	.18309	.98309	.18624	.3694	.0172	.4617	27
34	.18338	.98304	.18654	.3607	.0172	.4532	26
35	.18366	.98299	.18684	5.3521	1.0173	5.4447	25
36	.18395	.98293	.18714	.3434	.0174	.4362	24
37	.18424	.98288	.18745	.3349	.0174	.4278	23
38	.18452	.98283	.18775	.3263	.0175	.4194	22
39	.18481	.98277	.18805	.3178	.0175	.4110	21
40	.18509	.98272	.18835	5.3093	1.0176	5.4026	20
41	.18538	.98267	.18865	.3008	.0176	.3943	19
42	.18567	.98261	.18895	.2923	.0177	.3860	18
43	.18595	.98256	.18925	.2839	.0177	.3777	17
44	.18624	.98250	.18955	.2755	.0178	.3695	16
45	.18652	.98245	.18985	5.2671	1.0179	5.3612	15
46	.18681	.98240	.19016	.2588	.0179	.3530	14
47	.18709	.98234	.19046	.2505	.0180	.3449	13
48	.18738	.98229	.19076	.2422	.0180	.3367	12
49	.18767	.98223	.19106	.2339	.0181	.3286	11
50	.18795	.98218	.19136	5.2257	1.0181	5.3205	10
51	.18824	.98212	.19166	.2174	.0182	.3124	9
52	.18852	.98207	.19197	.2092	.0182	.3044	8
53	.18881	.98201	.19227	.2011	.0183	.2963	7
54	.18909	.98196	.19257	.1929	.0184	.2883	6
55	.18938	.98190	.19287	5.1848	1.0184	5.2803	5
56	.18967	.98185	.19317	.1767	.0185	.2724	4
57	.18995	.98179	.19347	.1686	.0185	.2645	3
58	.19024	.98174	.19378	.1606	.0186	.2566	2
59	.19052	.98168	.19408	.1525	.0186	.2487	1
60	.19081	.98163	.19438	5.1445	1.0187	5.2408	0
M	Cosine	Sine	Cotan.	Tan.	Cosec.	Secant	M

79°

11°

M	Sine	Cosine	Tan.	Cotan.	Secant	Cosec.	M
0	.19081	.98163	.19438	5.1445	1.0187	5.2408	60
1	.19109	.98157	.19468	.1366	.0188	.2330	59
2	.19138	.98152	.19498	.1286	.0188	.2252	58
3	.19166	.98146	.19529	.1207	.0189	.2174	57
4	.19195	.98140	.19559	.1128	.0189	.2097	56
5	.19224	.98135	.19589	5.1049	1.0190	5.2019	55
6	.19252	.98129	.19619	.0970	.0191	.1942	54
7	.19281	.98124	.19649	.0892	.0191	.1865	53
8	.19309	.98118	.19680	.0814	.0192	.1788	52
9	.19338	.98112	.19710	.0736	.0192	.1712	51
10	.19366	.98107	.19740	5.0658	1.0193	5.1636	50
11	.19395	.98101	.19770	.0581	.0193	.1560	49
12	.19423	.98095	.19800	.0504	.0194	.1484	48
13	.19452	.98090	.19831	.0427	.0195	.1409	47
14	.19480	.98084	.19861	.0350	.0195	.1333	46
15	.19509	.98078	.19891	5.0273	1.0196	5.1258	45
16	.19537	.98073	.19921	.0197	.0196	.1183	44
17	.19566	.98067	.19952	.0121	.0197	.1109	43
18	.19595	.98061	.19982	.0045	.0198	.1034	42
19	.19623	.98056	.20012	4.9969	.0198	.0960	41
20	.19652	.98050	.20042	4.9894	1.0199	5.0886	40
21	.19680	.98044	.20073	.9819	.0199	.0812	39
22	.19709	.98039	.20103	.9744	.0200	.0739	38
23	.19737	.98033	.20133	.9669	.0201	.0666	37
24	.19766	.98027	.20163	.9594	.0201	.0593	36
25	.19794	.98021	.20194	4.9520	1.0202	5.0520	35
26	.19823	.98016	.20224	.9446	.0202	.0447	34
27	.19851	.98010	.20254	.9372	.0203	.0375	33
28	.19880	.98004	.20285	.9298	.0204	.0302	32
29	.19908	.97998	.20315	.9225	.0204	.0230	31
30	.19937	.97992	.20345	4.9151	1.0205	5.0158	30
31	.19965	.97987	.20375	.9078	.0205	.0087	29
32	.19994	.97981	.20406	.9006	.0206	.0015	28
33	.20022	.97975	.20436	.8933	.0207	4.9944	27
34	.20051	.97969	.20466	.8860	.0207	.9873	26
35	.20079	.97963	.20497	4.8788	1.0208	4.9802	25
36	.20108	.97957	.20527	.8716	.0208	.9732	24
37	.20136	.97952	.20557	.8644	.0209	.9661	23
38	.20165	.97946	.20588	.8573	.0210	.9591	22
39	.20193	.97940	.20618	.8501	.0210	.9521	21
40	.20222	.97934	.20648	4.8430	1.0211	4.9452	20
41	.20250	.97928	.20679	.8359	.0211	.9382	19
42	.20279	.97922	.20709	.8288	.0212	.9313	18
43	.20307	.97916	.20739	.8217	.0213	.9243	17
44	.20336	.97910	.20770	.8147	.0213	.9175	16
45	.20364	.97904	.20800	4.8077	1.0214	4.9106	15
46	.20393	.97899	.20830	.8007	.0215	.9037	14
47	.20421	.97893	.20861	.7937	.0215	.8969	13
48	.20450	.97887	.20891	.7867	.0216	.8901	12
49	.20478	.97881	.20921	.7798	.0216	.8833	11
50	.20506	.97875	.20952	4.7728	1.0217	4.8765	10
51	.20535	.97869	.20982	.7659	.0218	.8697	9
52	.20563	.97863	.21012	.7591	.0218	.8630	8
53	.20592	.97857	.21043	.7522	.0219	.8563	7
54	.20620	.97851	.21073	.7453	.0220	.8496	6
55	.20649	.97845	.21104	4.7385	1.0220	4.8429	5
56	.20677	.97839	.21134	.7317	.0221	.8362	4
57	.20706	.97833	.21164	.7249	.0221	.8296	3
58	.20734	.97827	.21195	.7181	.0222	.8229	2
59	.20763	.97821	.21225	.7114	.0223	.8163	1
60	.20791	.97815	.21256	4.7046	1.0223	4.8097	0

M	Cosine	Sine	Cotan.	Tan.	Cosec.	Secant	M

78°

12°

M	Sine	Cosine	Tan.	Cotan.	Secant	Cosec.	M
0	.20791	.97815	.21256	4.7046	1.0223	4.8097	60
1	.20820	.97809	.21286	.6979	.0224	.8032	59
2	.20848	.97803	.21316	.6912	.0225	.7966	58
3	.20876	.97797	.21347	.6845	.0225	.7901	57
4	.20905	.97790	.21377	.6778	.0226	.7835	56
5	.20933	.97784	.21408	4.6712	1.0226	4.7770	55
6	.20962	.97778	.21438	.6646	.0227	.7706	54
7	.20990	.97772	.21468	.6580	.0228	.7641	53
8	.21019	.97766	.21499	.6514	.0228	.7576	52
9	.21047	.97760	.21529	.6448	.0229	.7512	51
10	.21076	.97754	.21560	4.6382	1.0230	4.7448	50
11	.21104	.97748	.21590	.6317	.0230	.7384	49
12	.21132	.97741	.21621	.6252	.0231	.7320	48
13	.21161	.97735	.21651	.6187	.0232	.7257	47
14	.21189	.97729	.21682	.6122	.0232	.7193	46
15	.21218	.97723	.21712	4.6057	1.0233	4.7130	45
16	.21246	.97717	.21742	.5993	.0234	.7067	44
17	.21275	.97711	.21773	.5928	.0234	.7004	43
18	.21303	.97704	.21803	.5864	.0235	.6942	42
19	.21331	.97698	.21834	.5800	.0235	.6879	41
20	.21360	.97692	.21864	4.5736	1.0236	4.6817	40
21	.21388	.97686	.21895	.5673	.0237	.6754	39
22	.21417	.97680	.21925	.5609	.0237	.6692	38
23	.21445	.97673	.21956	.5546	.0238	.6631	37
24	.21473	.97667	.21986	.5483	.0239	.6569	36
25	.21502	.97661	.22017	4.5420	1.0239	4.6507	35
26	.21530	.97655	.22047	.5357	.0240	.6446	34
27	.21559	.97648	.22078	.5294	.0241	.6385	33
28	.21587	.97642	.22108	.5232	.0241	.6324	32
29	.21615	.97636	.22139	.5169	.0242	.6263	31
30	.21644	.97630	.22169	4.5107	1.0243	4.6201	30
31	.21672	.97623	.22200	.5045	.0243	.6142	29
32	.21701	.97617	.22230	.4983	.0244	.6081	28
33	.21729	.97611	.22261	.4921	.0245	.6021	27
34	.21757	.97604	.22291	.4860	.0245	.5961	26
35	.21786	.97598	.22322	4.4799	1.0246	4.5901	25
36	.21814	.97592	.22353	.4737	.0247	.5841	24
37	.21843	.97585	.22383	.4676	.0247	.5782	23
38	.21871	.97579	.22414	.4615	.0248	.5722	22
39	.21899	.97573	.22444	.4555	.0249	.5663	21
40	.21928	.97566	.22475	4.4494	1.0249	4.5604	20
41	.21956	.97560	.22505	.4434	.0250	.5545	19
42	.21985	.97553	.22536	.4373	.0251	.5486	18
43	.22013	.97547	.22566	.4313	.0251	.5428	17
44	.22041	.97541	.22597	.4253	.0252	.5369	16
45	.22070	.97534	.22628	4.4194	1.0253	4.5311	15
46	.22098	.97528	.22658	.4134	.0253	.5253	14
47	.22126	.97521	.22689	.4075	.0254	.5195	13
48	.22155	.97515	.22719	.4015	.0255	.5137	12
49	.22183	.97508	.22750	.3956	.0255	.5079	11
50	.22211	.97502	.22781	4.3897	1.0256	4.5021	10
51	.22240	.97495	.22811	.3838	.0257	.4964	9
52	.22268	.97489	.22842	.3779	.0257	.4907	8
53	.22297	.97483	.22872	.3721	.0258	.4850	7
54	.22325	.97476	.22903	.3662	.0259	.4793	6
55	.22353	.97470	.22934	4.3604	1.0260	4.4736	5
56	.22382	.97463	.22964	.3546	.0260	.4679	4
57	.22410	.97457	.22995	.3488	.0261	.4623	3
58	.22438	.97450	.23025	.3430	.0262	.4566	2
59	.22467	.97443	.23056	.3372	.0262	.4510	1
60	.22495	.97437	.23087	4.3315	1.0263	4.4454	0

| M | Cosine | Sine | Cotan. | Tan. | Cosec. | Secant | M |

77°

41

13°

M	Sine	Cosine	Tan.	Cotan.	Secant	Cosec.	M
0	.22495	.97437	.23087	4.3315	1.0263	4.4454	60
1	.22523	.97430	.23117	.3257	.0264	.4398	59
2	.22552	.97424	.23148	.3200	.0264	.4342	58
3	.22580	.97417	.23179	.3143	.0265	.4287	57
4	.22608	.97411	.23209	.3086	.0266	.4231	56
5	.22637	.97404	.23240	4.3029	1.0266	4.4176	55
6	.22665	.97398	.23270	.2972	.0267	.4121	54
7	.22693	.97391	.23301	.2916	.0268	.4065	53
8	.22722	.97384	.23332	.2859	.0268	.4011	52
9	.22750	.97378	.23363	.2803	.0269	.3956	51
10	.22778	.97371	.23393	4.2747	1.0270	4.3901	50
11	.22807	.97364	.23424	.2691	.0271	.3847	49
12	.22835	.97358	.23455	.2635	.0271	.3792	48
13	.22863	.97351	.23485	.2579	.0272	.3738	47
14	.22892	.97344	.23516	.2524	.0273	.3684	46
15	.22920	.97338	.23547	4.2468	1.0273	4.3630	45
16	.22948	.97331	.23577	.2413	.0274	.3576	44
17	.22977	.97324	.23608	.2358	.0275	.3522	43
18	.23005	.97318	.23639	.2303	.0276	.3469	42
19	.23033	.97311	.23670	.2248	.0276	.3415	41
20	.23061	.97304	.23700	4.2193	1.0277	4.3362	40
21	.23090	.97298	.23731	.2139	.0278	.3309	39
22	.23118	.97291	.23752	.2084	.0278	.3256	38
23	.23146	.97284	.23793	.2030	.0279	.3203	37
24	.23175	.97277	.23823	.1976	.0280	.3150	36
25	.23203	.97271	.23854	4.1921	1.0280	4.3098	35
26	.23231	.97264	.23885	.1867	.0281	.3045	34
27	.23260	.97257	.23916	.1814	.0282	.2993	33
28	.23288	.97250	.23946	.1760	.0283	.2941	32
29	.23316	.97244	.23977	.1706	.0283	.2888	31
30	.23344	.97237	.24008	4.1653	1.0284	4.2836	30
31	.23373	.97230	.24039	.1600	.0285	.2785	29
32	.23401	.97223	.24069	.1546	.0285	.2733	28
33	.23429	.97216	.24100	.1493	.0286	.2681	27
34	.23458	.97210	.24131	.1440	.0287	.2630	26
35	.23486	.97203	.24162	4.1388	1.0288	4.2579	25
36	.23514	.97196	.24192	.1335	.0289	.2527	24
37	.23542	.97189	.24223	.1282	.0289	.2476	23
38	.23571	.97182	.24254	.1230	.0290	.2425	22
39	.23599	.97175	.24285	.1178	.0291	.2375	21
40	.23627	.97169	.24316	4.1126	1.0291	4.2324	20
41	.23655	.97162	.24346	.1073	.0292	.2273	19
42	.23684	.97155	.24377	.1022	.0293	.2223	18
43	.23712	.97148	.24408	.0970	.0293	.2173	17
44	.23740	.97141	.24439	.0918	.0294	.2122	16
45	.23768	.97134	.24470	4.0867	1.0295	4.2072	15
46	.23797	.97127	.24501	.0815	.0296	.2022	14
47	.23825	.97120	.24531	.0764	.0296	.1972	13
48	.23853	.97113	.24562	.0713	.0297	.1923	12
49	.23881	.97106	.24593	.0662	.0298	.1873	11
50	.23910	.97099	.24624	4.0611	1.0299	4.1824	10
51	.23938	.97092	.24655	.0560	.0299	.1774	9
52	.23966	.97086	.24686	.0509	.0300	.1725	8
53	.23994	.97079	.24717	.0458	.0301	.1676	7
54	.24023	.97072	.24747	.0408	.0302	.1627	6
55	.24051	.97065	.24778	4.0358	1.0302	4.1578	5
56	.24079	.97058	.24809	.0307	.0303	.1529	4
57	.24107	.97051	.24840	.0257	.0304	.1481	3
58	.24136	.97044	.24871	.0207	.0305	.1432	2
59	.24164	.97037	.24902	.0157	.0305	.1384	1
60	.24192	.97029	.24933	4.0108	1.0306	4.1336	0

M	Cosine	Sine	Cotan.	Tan.	Cosec.	Secant	M

76°

42

14°

M	Sine	Cosine	Tan.	Cotan.	Secant	Cosec.	M
0	.24192	.97029	.24933	4.0108	1.0306	4.1336	60
1	.24220	.97022	.24964	.0058	.0307	.1287	59
2	.24249	.97015	.24995	.0009	.0308	.1239	58
3	.24277	.97008	.25025	3.9959	.0308	.1191	57
4	.24305	.97001	.25056	.9910	.0309	.1144	56
5	.24333	.96994	.25087	3.9861	1.0310	4.1096	55
6	.24361	.96987	.25118	.9812	.0311	.1048	54
7	.24390	.96980	.25149	.9764	.0311	.1001	53
8	.24418	.96973	.25180	.9714	.0312	.0953	52
9	.24446	.96966	.25211	.9665	.0313	.0906	51
10	.24474	.96959	.25242	3.9616	1.0314	4.0859	50
11	.24502	.96952	.25273	.9568	.0314	.0812	49
12	.24531	.96944	.25304	.9520	.0315	.0765	48
13	.24559	.96937	.25335	.9471	.0316	.0718	47
14	.24587	.96930	.25366	.9423	.0317	.0672	46
15	.24615	.96923	.25397	3.9375	1.0317	4.0625	45
16	.24643	.96916	.25428	.9327	.0318	.0579	44
17	.24672	.96909	.25459	.9279	.0319	.0532	43
18	.24700	.96901	.25490	.9231	.0320	.0486	42
19	.24728	.96894	.25521	.9184	.0320	.0440	41
20	.24756	.96887	.25552	3.9136	1.0321	4.0394	40
21	.24784	.96880	.25583	.9089	.0322	.0348	39
22	.24813	.96873	.25614	.9042	.0323	.0302	38
23	.24841	.96865	.25645	.8994	.0323	.0256	37
24	.24869	.96858	.25676	.8947	.0324	.0211	36
25	.24897	.96851	.25707	3.8900	1.0325	4.0165	35
26	.24925	.96844	.25738	.8853	.0326	.0120	34
27	.24953	.96836	.25769	.8807	.0327	.0074	33
28	.24982	.96829	.25800	.8760	.0327	.0029	32
29	.25010	.96822	.25831	.8713	.0328	3.9984	31
30	.25038	.96815	.25862	3.8667	1.0329	3.9939	30
31	.25066	.96807	.25893	.8621	.0330	.9894	29
32	.25094	.96800	.25924	.8574	.0330	.9850	28
33	.25122	.96793	.25955	.8528	.0331	.9805	27
34	.25151	.96785	.25986	.8482	.0332	.9760	26
35	.25179	.96778	.26017	3.8436	1.0333	3.9716	25
36	.25207	.96771	.26048	.8390	.0334	.9672	24
37	.25235	.96763	.26079	.8345	.0334	.9627	23
38	.25263	.96756	.26110	.8299	.0335	.9583	22
39	.25291	.96749	.26141	.8254	.0336	.9539	21
40	.25319	.96741	.26172	3.8208	1.0337	3.9495	20
41	.25348	.96734	.26203	.8163	.0338	.9451	19
42	.25376	.96727	.26234	.8118	.0338	.9408	18
43	.25404	.96719	.26266	.8073	.0339	.9364	17
44	.25432	.96712	.26297	.8027	.0340	.9320	16
45	.25460	.96704	.26328	3.7983	1.0341	3.9277	15
46	.25488	.96697	.26359	.7938	.0341	.9234	14
47	.25516	.96690	.26390	.7893	.0342	.9190	13
48	.25544	.96682	.26421	.7848	.0343	.9147	12
49	.25573	.96675	.26452	.7804	.0344	.9104	11
50	.25601	.96667	.26483	3.7759	1.0345	3.9061	10
51	.25629	.96660	.26514	.7715	.0345	.9018	9
52	.25657	.96652	.26546	.7671	.0346	.8976	8
53	.25685	.96645	.26577	.7627	.0347	.8933	7
54	.25713	.96638	.26608	.7583	.0348	.8890	6
55	.25741	.96630	.26639	3.7539	1.0349	3.8848	5
56	.25769	.96623	.26670	.7495	.0349	.8805	4
57	.25798	.96615	.26701	.7451	.0350	.8763	3
58	.25826	.96608	.26732	.7407	.0351	.8721	2
59	.25854	.96600	.26764	.7364	.0352	.8679	1
60	.25882	.96592	.26795	3.7320	1.0353	3.8637	0

M	Cosine	Sine	Cotan.	Tan.	Cosec.	Secant	M

75°

43

15°

M	Sine	Cosine	Tan.	Cotsn.	Secant	Cosec.	M
0	.25882	.96592	.26795	3.7320	1.0353	3.8637	60
1	.25910	.96585	.26826	.7277	.0353	.8595	59
2	.25938	.96577	.26857	.7234	.0354	.8553	58
3	.25966	.96570	.26888	.7191	.0355	.8512	57
4	.25994	.96562	.26920	.7147	.0356	.8470	56
5	.26022	.96555	.26951	3.7104	1.0357	3.8428	55
6	.26050	.96547	.26982	.7062	.0358	.8387	54
7	.26078	.96540	.27013	.7019	.0358	.8346	53
8	.26107	.96532	.27044	.6976	.0359	.8304	52
9	.26135	.96524	.27076	.6933	.0360	.8263	51
10	.26163	.96517	.27107	3.6891	1.0361	3.8222	50
11	.26191	.96509	.27138	.6848	.0362	.8181	49
12	.26219	.96502	.27169	.6806	.0362	.8140	48
13	.26247	.96494	.27201	.6764	.0363	.8100	47
14	.26275	.96486	.27232	.6722	.0364	.8059	46
15	.26303	.96479	.27263	3.6679	1.0365	3.8018	45
16	.26331	.96471	.27294	.6637	.0366	.7978	44
17	.26359	.96463	.27326	.6596	.0367	.7937	43
18	.26387	.96456	.27357	.6554	.0367	.7897	42
19	.26415	.96448	.27388	.6512	.0368	.7857	41
20	.26443	.96440	.27419	3.6470	1.0369	3.7816	40
21	.26471	.96433	.27451	.6429	.0370	.7776	39
22	.26499	.96425	.27482	.6387	.0371	.7736	38
23	.26527	.96417	.27513	.6346	.0371	.7697	37
24	.26556	.96409	.27544	.6305	.0372	.7657	36
25	.26584	.96402	.27576	3.6263	1.0373	3.7617	35
26	.26612	.96394	.27607	.6222	.0374	.7577	34
27	.26640	.96386	.27638	.6181	.0375	.7538	33
28	.26668	.96378	.27670	.6140	.0376	.7498	32
29	.26696	.96371	.27701	.6100	.0376	.7459	31
30	.26724	.96363	.27732	3.6059	1.0377	3.7420	30
31	.26752	.96355	.27764	.6018	.0378	.7380	29
32	.26780	.96347	.27795	.5977	.0379	.7341	28
33	.26808	.96340	.27826	.5937	.0380	.7302	27
34	.26836	.96332	.27858	.5896	.0381	.7263	26
35	.26864	.96324	.27889	3.5856	1.0382	3.7224	25
36	.26892	.96316	.27920	.5816	.0382	.7186	24
37	.26920	.96308	.27952	.5776	.0383	.7147	23
38	.26948	.96301	.27983	.5736	.0384	.7108	22
39	.26976	.96293	.28014	.5696	.0385	.7070	21
40	.27004	.96285	.28046	3.5656	1.0386	3.7031	20
41	.27032	.96277	.28077	.5616	.0387	.6993	19
42	.27060	.96269	.28109	.5576	.0387	.6955	18
43	.27088	.96261	.28140	.5536	.0388	.6917	17
44	.27116	.96253	.28171	.5497	.0389	.6878	16
45	.27144	.96245	.28203	3.5457	1.0390	3.6840	15
46	.27172	.96238	.28234	.5418	.0391	.6802	14
47	.27200	.96230	.28266	.5378	.0392	.6765	13
48	.27228	.96222	.28297	.5339	.0393	.6727	12
49	.27256	.96214	.28328	.5300	.0393	.6689	11
50	.27284	.96206	.28360	3.5261	1.0394	3.6651	10
51	.27312	.96198	.28391	.5222	.0395	.6614	9
52	.27340	.96190	.28423	.5183	.0396	.6576	8
53	.27368	.96182	.28454	.5144	1.0397	.6539	7
54	.27396	.96174	.28486	.5105	.0398	.6502	6
55	.27424	.96166	.28517	3.5066	1.0399	3.6464	5
56	.27452	.96158	.28549	.5028	.0399	.6427	4
57	.27480	.96150	.28580	.4989	.0400	.6390	3
58	.27508	.96142	.28611	.4951	.0401	.6353	2
59	.27536	.96134	.28643	.4912	.0402	.6316	1
60	.27564	.96126	.28674	3.4874	1.0403	3.6279	0

M	Cosine	Sine	Cotan.	Tan.	Cosec.	Secant	M

74°

16°

M	Sine	Cosine	Tan.	Cotan.	Secant	Cosec.	M
0	.27564	.96126	.28674	3.4874	1.0403	3.6279	60
1	.27592	.96118	.28706	.4836	.0404	.6243	69
2	.27620	.96110	.28737	.4798	.0405	.6206	58
3	.27648	.96102	.28769	.4760	.0406	.6169	57
4	.27675	.96094	.28800	.4722	.0406	.6133	56
5	.27703	.96086	.28832	3.4684	1.0407	3.6096	55
6	.27731	.96078	.28863	.4646	.0408	.6060	54
7	.27759	.96070	.28895	.4608	.0409	.6024	53
8	.27787	.96062	.28926	.4570	.0410	.5987	52
9	.27815	.96054	.28958	.4533	.0411	.5951	51
10	.27843	.96045	.28990	3.4495	1.0412	3.5915	50
11	.27871	.96037	.29021	.4458	.0413	.5879	49
12	.27899	.96029	.29053	.4420	.0413	.5843	48
13	.27927	.96021	.29084	.4383	.0414	.5807	47
14	.27955	.96013	.29116	.4346	.0415	.5772	46
15	.27983	.96005	.29147	3.4308	1.0416	3.5736	45
16	.28011	.95997	.29179	.4271	.0417	.5700	44
17	.28039	.95989	.29210	.4234	.0418	.5665	43
18	.28067	.95980	.29242	.4197	.0419	.5629	42
19	.28094	.95972	.29274	.4160	.0420	.5594	41
20	.28122	.95964	.29305	3.4124	1.0420	3.5559	40
21	.28150	.95956	.29337	.4087	.0421	.5523	39
22	.28178	.95948	.29368	.4050	.0422	.5488	38
23	.28206	.95940	.29400	.4014	.0423	.5453	37
24	.28234	.95931	.29432	.3977	.0424	.5418	36
25	.28262	.95923	.29463	3.3941	1.0425	3.5383	35
26	.28290	.95915	.29495	.3904	.0426	.5348	34
27	.28318	.95907	.29526	.3868	.0427	.5313	33
28	.28346	.95898	.29558	.3832	.0428	.5279	32
29	.28374	.95890	.29590	.3795	.0428	.5244	31
30	.28401	.95882	.29621	3.3759	1.0429	3.5209	30
31	.28429	.95874	.29653	.3723	.0430	.5175	29
32	.28457	.95865	.29685	.3687	.0431	.5140	28
33	.28485	.95857	.29716	.3651	.0432	.5106	27
34	.28513	.95849	.29748	.3616	.0433	.5072	26
35	.28541	.95840	.29780	3.3580	1.0434	3.5037	25
36	.28569	.95832	.29811	.3544	.0435	.5003	24
37	.28597	.95824	.29843	.3509	.0436	.4969	23
38	.28624	.95816	.29875	.3473	.0437	.4935	22
39	.28652	.95807	.29906	.3438	.0438	.4901	21
40	.28680	.95799	.29938	3.3402	1.0438	3.4867	20
41	.28708	.95791	.29970	.3367	.0439	.4833	19
42	.28736	.95782	.30001	.3332	.0440	.4799	18
43	.28764	.95774	.30033	.3296	.0441	.4766	17
44	.28792	.95765	.30065	.3261	.0442	.4732	16
45	.28820	.95757	.30096	3.3226	1.0443	3.4698	15
46	.28847	.95749	.30128	.3191	.0444	.4665	14
47	.28875	.95740	.30160	.3156	.0445	.4632	13
48	.28903	.95732	.30192	.3121	.0446	.4598	12
49	.28931	.95723	.30223	.3087	.0447	.4565	11
50	.28959	.95715	.30255	3.3052	1.0448	3.4532	10
51	.28987	.95707	.30287	.3017	.0448	.4498	9
52	.29014	.95698	.30319	.2983	.0449	.4465	8
53	.29042	.95690	.30350	.2948	.0450	.4432	7
54	.29070	.95681	.30382	.2914	.0451	.4399	6
55	.29098	.95673	.30414	3.2879	1.0452	3.4366	5
56	.29126	.95664	.30446	.2845	.0453	.4334	4
57	.29154	.95656	.30478	.2811	.0454	.4301	3
58	.29181	.95647	.30509	.2777	.0455	.4268	2
59	.29209	.95639	.30541	.2742	.0456	.4236	1
60	.29237	.95630	.30573	.2708	1.0457	3.4203	0
M	Cosine	Sine	Cotan.	Tan.	Cosec.	Secant	M

73°

45

17°

M	Sine	Cosine	Tan.	Cotan.	Secant	Cosec.	M
0	.29237	.95630	.30573	3.2708	1.0457	3.4203	60
1	.29265	.95622	.30605	.2674	.0458	.4170	59
2	.29293	.95613	.30637	.2640	.0459	.4138	58
3	.29321	.95605	.30668	.2607	.0460	.4106	57
4	.29348	.95596	.30700	.2573	.0461	.4073	56
5	.29376	.95588	.30732	3.2539	1.0461	3.4041	55
6	.29404	.95579	.30764	.2505	.0462	.4009	54
7	.29432	.95571	.30796	.2472	.0463	.3977	53
8	.29460	.95562	.30828	.2438	.0464	.3945	52
9	.29487	.95554	.30859	.2405	.0465	.3913	51
10	.29515	.95545	.30891	3.2371	1.0466	3.3881	50
11	.29543	.95536	.30923	.2338	.0467	.3849	49
12	.29571	.95528	.30955	.2305	.0468	.3817	48
13	.29598	.95519	.30987	.2271	.0469	.3785	47
14	.29626	.95511	.31019	.2238	.0470	.3754	46
15	.29654	.95502	.31051	3.2205	1.0471	3.3722	45
16	.29682	.95493	.31083	.2172	.0472	.3690	44
17	.29710	.95485	.31115	.2139	.0473	.3659	43
18	.29737	.95476	.31146	.2106	.0474	.3627	42
19	.29765	.95467	.31178	.2073	.0475	.3596	41
20	.29793	.95459	.31210	3.2041	1.0476	3.3565	40
21	.29821	.95450	.31242	.2008	.0477	.3534	39
22	.29848	.95441	.31274	.1975	.0478	.3502	38
23	.29876	.95433	.31306	.1942	.0478	.3471	37
24	.29904	.95424	.31338	.1910	.0479	.3440	36
25	.29932	.95415	.31370	3.1877	1.0480	3.3409	35
26	.29959	.95407	.31402	.1845	.0481	.3378	34
27	.29987	.95398	.31434	.1813	.0482	.3347	33
28	.30015	.95389	.31466	.1780	.0483	.3316	32
29	.30043	.95380	.31498	.1748	.0484	.3286	31
30	.30070	.95372	.31530	3.1716	1.0485	3.3255	30
31	.30098	.95363	.31562	.1684	.0486	.3224	29
32	.30126	.95354	.31594	.1652	.0487	.3194	28
33	.30154	.95345	.31626	.1620	.0488	.3163	27
34	.30181	.95337	.31658	.1588	.0489	.3133	26
35	.30209	.95328	.31690	3.1556	1.0490	3.3102	25
36	.30237	.95319	.31722	.1524	.0491	.3072	24
37	.30265	.95310	.31754	.1492	.0492	.3042	23
38	.30292	.95301	.31786	.1460	.0493	.3011	22
39	.30320	.95293	.31818	.1429	.0494	.2981	21
40	.30348	.95284	.31850	3.1397	1.0495	3.2951	20
41	.30375	.95275	.31882	.1366	.0496	.2921	19
42	.30403	.95266	.31914	.1334	.0497	.2891	18
43	.30431	.95257	.31946	.1303	.0498	.2861	17
44	.30459	.95248	.31978	.1271	.0499	.2831	16
45	.30486	.95239	.32010	3.1240	1.0500	3.2801	15
46	.30514	.95231	.32042	.1209	.0501	.2772	14
47	.30542	.95222	.32074	.1177	.0502	.2742	13
48	.30569	.95213	.32106	.1146	.0503	.2712	12
49	.30597	.95204	.32138	.1115	.0504	.2683	11
50	.30625	.95195	.32171	3.1084	1.0505	3.2653	10
51	.30653	.95186	.32203	.1053	.0506	.2624	9
52	.30680	.95177	.32235	.1022	.0507	.2594	8
53	.30708	.95168	.32267	.0991	.0508	.2565	7
54	.30736	.95159	.32299	.0960	.0509	.2535	6
55	.30763	.95150	.32331	3.0930	1.0510	3.2506	5
56	.30791	.95141	.32363	.0899	.0511	.2477	4
57	.30819	.95132	.32395	.0868	.0512	.2448	3
58	.30846	.95124	.32428	.0838	.0513	.2419	2
59	.30874	.95115	.32460	.0807	.0514	.2390	1
60	.30902	.95106	.32492	3.0777	1.0515	3.2361	0
M	Cosine	Sine	Cotan.	Tan.	Cosec.	Secant	M

72°

46

18°

M	Sine	Cosine	Tan.	Cotan.	Secant	Cosec.	M
0	.30902	.95106	.32492	3.0777	1.0515	3.2361	60
1	.30929	.95097	.32524	.0746	.0516	.2332	59
2	.30957	.95088	.32556	.0716	.0517	.2303	58
3	.30985	.95079	.32588	.0686	.0518	.2274	57
4	.31012	.95070	.32621	.0655	.0519	.2245	56
5	.31040	.95061	.32653	3.0625	1.0520	3.2216	55
6	.31068	.95051	.32685	.0595	.0521	.2188	54
7	.31095	.95042	.32717	.0565	.0522	.2159	53
8	.31123	.95033	.32749	.0535	.0523	.2131	52
9	.31150	.95024	.32782	.0505	.0524	.2102	51
10	.31178	.95015	.32814	3.0475	1.0525	3.2074	50
11	.31206	.95006	.32846	.0445	.0526	.2045	49
12	.31233	.94997	.32878	.0415	.0527	.2017	48
13	.31261	.94988	.32910	.0385	.0528	.1989	47
14	.31289	.94979	.32943	.0356	.0529	.1960	46
15	.31316	.94970	.32975	3.0326	1.0530	3.1932	45
16	.31344	.94961	.33007	.0296	.0531	.1904	44
17	.31372	.94952	.33039	.0267	.0532	.1876	43
18	.31399	.94942	.33072	.0237	.0533	.1848	42
19	.31427	.94933	.33104	.0208	.0534	.1820	41
20	.31454	.94924	.33136	3.0178	1.0535	3.1792	40
21	.31482	.94915	.33169	.0149	.0536	.1764	39
22	.31510	.94906	.33201	.0120	.0537	.1736	38
23	.31537	.94897	.33233	.0090	.0538	.1708	37
24	.31565	.94888	.33265	.0061	.0539	.1681	36
25	.31592	.94878	.33298	3.0032	1.0540	3.1653	35
26	.31620	.94869	.33330	.0003	.0541	.1625	34
27	.31648	.94860	.33362	2.9974	.0542	.1598	33
28	.31675	.94851	.33395	.9945	.0543	.1570	32
29	.31703	.94841	.33427	.9916	.0544	.1543	31
30	.31730	.94832	.33459	2.9887	1.0545	3.1515	30
31	.31758	.94823	.33492	.9858	.0546	.1488	29
32	.31786	.94814	.33524	.9829	.0547	.1461	28
33	.31813	.94805	.33557	.9800	.0548	.1433	27
34	.31841	.94795	.33589	.9772	.0549	.1406	26
35	.31868	.94786	.33621	2.9743	1.0550	3.1379	25
36	.31896	.94777	.33654	.9714	.0551	.1352	24
37	.31923	.94767	.33686	.9686	.0552	.1325	23
38	.31951	.94758	.33718	.9657	.0553	.1298	22
39	.31978	.94749	.33751	.9629	.0554	.1271	21
40	.32006	.94740	.33783	2.9600	1.0555	3.1244	20
41	.32034	.94730	.33816	.9572	.0556	.1217	19
42	.32061	.94721	.33848	.9544	.0557	.1190	18
43	.32089	.94712	.33880	.9515	.0558	.1163	17
44	.32116	.94702	.33913	.9487	.0559	.1137	16
45	.32144	.94693	.33945	2.9459	1.0560	3.1110	15
46	.32171	.94684	.33978	.9431	.0561	.1083	14
47	.32199	.94674	.34010	.9403	.0562	.1057	13
48	.32226	.94665	.34043	.9375	.0563	.1030	12
49	.32254	.94655	.34075	.9347	.0565	.1004	11
50	.32282	.94646	.34108	2.9319	1.0566	3.0977	10
51	.32309	.94637	.34140	.9291	.0567	.0951	9
52	.32337	.94627	.34173	.9263	.0568	.0925	8
53	.32364	.94618	.34205	.9235	.0569	.0898	7
54	.32392	.94608	.34238	.9207	.0570	.0872	6
55	.32419	.94599	.34270	2.9180	1.0571	3.0846	5
56	.32447	.94590	.34303	.9152	.0572	.0820	4
57	.32474	.94580	.34335	.9125	.0573	.0793	3
58	.32502	.94571	.34368	.9097	.0574	.0767	2
59	.32529	.94561	.34400	.9069	.0575	.0741	1
60	.32557	.94552	.34433	2.9042	1.0576	3.0715	0

| M | Cosine | Sine | Cotan. | Tan. | Cosec. | Secant | M |

71°

47

19°

M	Sine	Cosine	Tan.	Cotan.	Secant	Cosec.	M
0	.32557	.94552	.34433	2.9042	1.0576	3.0715	60
1	.32584	.94542	.34465	.9015	.0577	.0690	59
2	.32612	.94533	.34498	.8987	.0578	.0664	58
3	.32639	.94523	.34530	.8960	.0579	.0638	57
4	.32667	.94514	.34563	.8933	.0580	.0612	56
5	.32694	.94504	.34595	2.8905	1.0581	3.0586	55
6	.32722	.94495	.34628	.8878	.0582	.0561	54
7	.32749	.94485	.34661	.8851	.0584	.0535	53
8	.32777	.94476	.34693	.8824	.0585	.0509	52
9	.32804	.94466	.34726	.8797	.0586	.0484	51
10	.32832	.94457	.34758	2.8770	1.0587	3.0458	50
11	.32859	.94447	.34791	.8743	.0588	.0433	49
12	.32887	.94438	.34824	.8716	.0589	.0407	48
13	.32914	.94428	.34856	.8689	.0590	.0382	47
14	.32942	.94418	.34889	.8662	.0591	.0357	46
15	.32969	.94409	.34921	2.8636	1.0592	3.0331	45
16	.32996	.94399	.34954	.8609	.0593	.0306	44
17	.33024	.94390	.34987	.8582	.0594	.0281	43
18	.33051	.94380	.35019	.8555	.0595	.0256	42
19	.33079	.94370	.35052	.8529	.0596	.0231	41
20	.33106	.94361	.35085	2.8502	1.0598	3.0206	40
21	.33134	.94351	.35117	.8476	.0599	.0181	39
22	.33161	.94341	.35150	.8449	.0600	.0156	38
23	.33189	.94332	.35183	.8423	.0601	.0131	37
24	.33216	.94322	.35215	.8396	.0602	.0106	36
25	.33243	.94313	.35248	2.8370	1.0603	3.0081	35
26	.33271	.94303	.35281	.8344	.0604	.0056	34
27	.33298	.94293	.35314	.8318	.0605	.0031	33
28	.33326	.94283	.35346	.8291	.0606	.0007	32
29	.33353	.94274	.35379	.8265	.0607	2.9982	31
30	.33381	.94264	.35412	2.8239	1.0608	2.9957	30
31	.33408	.94254	.35445	.8213	.0609	.9933	29
32	.33435	.94245	.35477	.8187	.0611	.9908	28
33	.33463	.94235	.35510	.8161	.0612	.9884	27
34	.33490	.94225	.35543	.8135	.0613	.9859	26
35	.33518	.94215	.35576	2.8109	1.0614	2.9835	25
36	.33545	.94206	.35608	.8083	.0615	.9810	24
37	.33572	.94196	.35641	.8057	.0616	.9786	23
38	.33600	.94186	.35674	.8032	.0617	.9762	22
39	.33627	.94176	.35707	.8006	.0618	.9738	21
40	.33655	.94167	.35739	2.7980	1.0619	2.9713	20
41	.33682	.94157	.35772	.7954	.0620	.9689	19
42	.33709	.94147	.35805	.7929	.0622	.9665	18
43	.33737	.94137	.35838	.7903	.0623	.9641	17
44	.33764	.94127	.35871	.7878	.0624	.9617	16
45	.33792	.94118	.35904	2.7852	1.0625	2.9593	15
46	.33819	.94108	.35936	.7827	.0626	.9569	14
47	.33846	.94098	.35969	.7801	.0627	.9545	13
48	.33874	.94088	.36002	.7776	.0628	.9521	12
49	.33901	.94078	.36035	.7751	.0629	.9497	11
50	.33928	.94068	.36068	2.7725	1.0630	2.9474	10
51	.33956	.94058	.36101	.7700	.0632	.9450	9
52	.33983	.94049	.36134	.7675	.0633	.9426	8
53	.34011	.94039	.36167	.7650	.0634	.9402	7
54	.34038	.94029	.36199	.7625	.0635	.9379	6
55	.34065	.94019	.36232	2.7600	1.0636	2.9355	5
56	.34093	.94009	.36265	.7575	.0637	.9332	4
57	.34120	.93999	.36298	.7550	.0638	.9308	3
58	.34147	.93989	.36331	.7525	.0639	.9285	2
59	.34175	.93979	.36364	.7500	.0641	.9261	1
60	.34202	.93969	.36397	2.7475	1.0642	2.9238	0

M	Cosine	Sine	Cotan.	Tan.	Cosec.	Secant	M

70°

M	Sine	Cosine	Tan.	Cotan.	Secant	Cosec.	M
0	.34202	.93969	.36397	2.7475	1.0642	2.9238	60
1	.34229	.93959	.36430	.7450	.0643	.9215	59
2	.34257	.93949	.36463	.7425	.0644	.9191	58
3	.34284	.93939	.36496	.7400	.0645	.9168	57
4	.34311	.93929	.36529	.7376	.0646	.9145	56
5	.34339	.93919	.36562	2.7351	1.0647	2.9122	55
6	.34366	.93909	.36595	.7326	.0648	.9098	54
7	.34393	.93899	.36628	.7302	.0650	.9075	53
8	.34421	.93889	.36661	.7277	.0651	.9052	52
9	.34448	.93879	.36694	.7252	.0652	.9029	51
10	.34475	.93869	.36727	2.7228	1.0653	2.9006	50
11	.34502	.93859	.36760	.7204	.0654	.8983	49
12	.34530	.93849	.36793	.7179	.0655	.8960	48
13	.34557	.93839	.36826	.7155	.0656	.8937	47
14	.34584	.93829	.36859	.7130	.0658	.8915	46
15	.34612	.93819	.36892	2.7106	1.0659	2.8892	45
16	.34639	.93809	.36925	.7082	.0660	.8869	44
17	.34666	.93799	.36958	.7058	.0661	.8846	43
18	.34693	.93789	.36991	.7033	.0662	.8824	42
19	.34721	.93779	.37024	.7009	.0663	.8801	41
20	.34748	.93769	.37057	2.6985	1.0664	2.8778	40
21	.34775	.93758	.37090	.6961	.0666	.8756	39
22	.34803	.93748	.37123	.6937	.0667	.8733	38
23	.34830	.93738	.37156	.6913	.0668	.8711	37
24	.34857	.93728	.37190	.6889	.0669	.8688	36
25	.34884	.93718	.37223	2.6865	1.0670	2.8666	35
26	.34912	.93708	.37256	.6841	.0671	.8644	34
27	.34939	.93698	.37289	.6817	.0673	.8621	33
28	.34966	.93687	.37322	.6794	.0674	.8599	32
29	.34993	.93677	.37355	.6770	.0675	.8577	31
30	.35021	.93667	.37388	2.6746	1.0676	2.8554	30
31	.35048	.93657	.37422	.6722	.0677	.8532	29
32	.35075	.93647	.37455	.6699	.0678	.8510	28
33	.35102	.93637	.37488	.6675	.0679	.8488	27
34	.35130	.93626	.37521	.6652	.0681	.8466	26
35	.35157	.93616	.37554	2.6628	1.0682	2.8444	25
36	.35184	.93606	.37587	.6604	.0683	.8422	24
37	.35211	.93596	.37621	.6581	.0684	.8400	23
38	.35239	.93585	.37654	.6558	.0685	.8378	22
39	.35266	.93575	.37687	.6534	.0686	.8356	21
40	.35293	.93565	.37720	2.6511	1.0688	2.8334	20
41	.35320	.93555	.37754	.6487	.0689	.8312	19
42	.35347	.93544	.37787	.6464	.0690	.8290	18
43	.35375	.93534	.37820	.6441	.0691	.8269	17
44	.35402	.93524	.37853	.6418	.0692	.8247	16
45	.35429	.93513	.37887	2.6394	1.0694	2.8225	15
46	.35456	.93503	.37920	.6371	.0695	.8204	14
47	.35483	.93493	.37953	.6348	.0696	.8182	13
48	.35511	.93482	.37986	.6325	.0697	.8160	12
49	.35538	.93472	.38020	.6302	.0698	.8139	11
50	.35565	.93462	.38053	2.6279	1.0699	2.8117	10
51	.35592	.93451	.38086	.6256	.0701	.8096	9
52	.35619	.93441	.38120	.6233	.0702	.8074	8
53	.35647	.93431	.38153	.6210	.0703	.8053	7
54	.35674	.93420	.38186	.6187	.0704	.8032	6
55	.35701	.93410	.38220	2.6164	1.0705	2.8010	5
56	.35728	.93400	.38253	.6142	.0707	.7989	4
57	.35755	.93389	.38286	.6119	.0708	.7968	3
58	.35782	.93379	.38320	.6096	.0709	.7947	2
59	.35810	.93368	.38353	.6073	.0710	.7925	1
60	.35837	.93358	.38386	.6051	1.0711	2.7904	0
M	Cosine	Sine	Cotan.	Tan.	Cosec.	Secant	M

21°

M	Sine	Cosine	Tan.	Cotan.	Secant	Cosec.	M
0	.35837	.93358	.38386	2.6051	1.0711	2.7904	60
1	.35864	.93348	.38420	.6028	.0713	.7883	59
2	.35891	.93337	.38453	.6006	.0714	.7862	58
3	.35918	.93327	.38486	.5983	.0715	.7841	57
4	.35945	.93316	.38520	.5960	.0716	.7820	56
5	.35972	.93306	.38553	2.5938	1.0717	2.7799	55
6	.36000	.93295	.38587	.5916	.0719	.7778	54
7	.36027	.93285	.38620	.5893	.0720	.7757	53
8	.36054	.93274	.38654	.5871	.0721	.7736	52
9	.36081	.93264	.38687	.5848	.0722	.7715	51
10	.36108	.93253	.38720	2.5826	1.0723	2.7694	50
11	.36135	.93243	.38754	.5804	.0725	.7674	49
12	.36162	.93232	.38787	.5781	.0726	.7653	48
13	.36189	.93222	.38821	.5759	.0727	.7632	47
14	.36217	.93211	.38854	.5737	.0728	.7611	46
15	.36244	.93201	.38888	2.5715	1.0729	2.7591	45
16	.36271	.93190	.38921	.5693	.0731	.7570	44
17	.36298	.93180	.38955	.5671	.0732	.7550	43
18	.36325	.93169	.38988	.5649	.0733	.7529	42
19	.36352	.93158	.39022	.5627	.0734	.7509	41
20	.36379	.93148	.39055	2.5605	1.0736	2.7488	40
21	.36406	.93137	.39089	.5583	.0737	.7468	39
22	.36433	.93127	.39122	.5561	.0738	.7447	38
23	.36460	.93116	.39156	.5539	.0739	.7427	37
24	.36488	.93105	.39189	.5517	.0740	.7406	36
25	.36515	.93095	.39223	2.5495	1.0742	2.7386	35
26	.36542	.93084	.39257	.5473	.0743	.7366	34
27	.36569	.93074	.39290	.5451	.0744	.7346	33
28	.36596	.93063	.39324	.5430	.0745	.7325	32
29	.36623	.93052	.39357	.5408	.0747	.7305	31
30	.36650	.93042	.39391	2.5386	1.0748	2.7285	30
31	.36677	.93031	.39425	.5365	.0749	.7265	29
32	.36704	.93020	.39458	.5343	.0750	.7245	28
33	.36731	.93010	.39492	.5322	.0751	.7225	27
34	.36758	.92999	.39525	.5300	.0753	.7205	26
35	.36785	.92988	.39559	2.5278	1.0754	2.7185	25
36	.36812	.92978	.39593	.5257	.0755	.7165	24
37	.36839	.92967	.39626	.5236	.0756	.7145	23
38	.36866	.92956	.39660	.5214	.0758	.7125	22
39	.36893	.92945	.39694	.5193	.0759	.7105	21
40	.36921	.92935	.39727	2.5171	1.0760	2.7085	20
41	.36948	.92924	.39761	.5150	.0761	.7065	19
42	.36975	.92913	.39795	.5129	.0763	.7045	18
43	.37002	.92902	.39828	.5108	.0764	.7026	17
44	.37029	.92892	.39862	.5086	.0765	.7006	16
45	.37056	.92881	.39896	2.5065	1.0766	2.6986	15
46	.37083	.92870	.39930	.5044	.0768	.6967	14
47	.37110	.92859	.39963	.5023	.0769	.6947	13
48	.37137	.92848	.39997	.5002	.0770	.6927	12
49	.37164	.92838	.40031	.4981	.0771	.6908	11
50	.37191	.92827	.40065	2.4960	1.0773	2.6888	10
51	.37218	.92816	.40098	.4939	.0774	.6869	9
52	.37245	.92805	.40132	.4918	.0775	.6849	8
53	.37272	.92794	.40166	.4897	.0776	.6830	7
54	.37299	.92784	.40200	.4876	.0778	.6810	6
55	.37326	.92773	.40233	2.4855	1.0779	2.6791	5
56	.37353	.92762	.40267	.4834	.0780	.6772	4
57	.37380	.92751	.40301	.4813	.0781	.6752	3
58	.37407	.92740	.40335	.4792	.0783	.6733	2
59	.37434	.92729	.40369	.4772	.0784	.6714	1
60	.37461	.92718	.40403	2.4751	1.0785	2.6695	0

| M | Cosine | Sine | Cotan. | Tan. | Cosec. | Secant | M |

68°

22°

M	Sine	Cosine	Tan.	Cotan.	Secant	Cosec.	M
0	.37461	.92718	.40403	2.4751	1.0785	2.6695	60
1	.37488	.92707	.40436	.4730	.0787	.6675	59
2	.37514	.92696	.40470	.4709	.0788	.6656	58
3	.37541	.92686	.40504	.4689	.0789	.6637	57
4	.37568	.92675	.40538	.4668	.0790	.6618	56
5	.37595	.92664	.40572	2.4647	1.0792	2.6599	55
6	.37622	.92653	.40606	.4627	.0793	.6580	54
7	.37649	.92642	.40640	.4606	.0794	.6561	53
8	.37676	.92631	.40673	.4586	.0795	.6542	52
9	.37703	.92620	.40707	.4565	.0797	.6523	51
10	.37730	.92609	.40741	2.4545	1.0798	2.6504	50
11	.37757	.92598	.40775	.4525	.0799	.6485	49
12	.37784	.92587	.40809	.4504	.0801	.6466	48
13	.37811	.92576	.40843	.4484	.0802	.6447	47
14	.37838	.92565	.40877	.4463	.0803	.6428	46
15	.37865	.92554	.40911	2.4443	1.0804	2.6410	45
16	.37892	.92543	.40945	.4423	.0806	.6391	44
17	.37919	.92532	.40979	.4403	.0807	.6372	43
18	.37946	.92521	.41013	.4382	.0808	.6353	42
19	.37972	.92510	.41047	.4362	.0810	.6335	41
20	.37999	.92499	.41081	2.4342	1.0811	2.6316	40
21	.38026	.92488	.41115	.4322	.0812	.6297	39
22	.38053	.92477	.41149	.4302	.0813	.6279	38
23	.38080	.92466	.41183	.4282	.0815	.6260	37
24	.38107	.92455	.41217	.4262	.0816	.6242	36
25	.38134	.92443	.41251	2.4242	1.0817	2.6223	35
26	.38161	.92432	.41285	.4222	.0819	.6205	34
27	.38188	.92421	.41319	.4202	.0820	.6186	33
28	.38214	.92410	.41353	.4182	.0821	.6168	32
29	.38241	.92399	.41387	.4162	.0823	.6150	31
30	.38268	.92388	.41421	2.4142	1.0824	2.6131	30
31	.38295	.92377	.41455	.4122	.0825	.6113	29
32	.38322	.92366	.41489	.4102	.0826	.6095	28
33	.38349	.92354	.41524	.4083	.0828	.6076	27
34	.38376	.92343	.41558	.4063	.0829	.6058	26
35	.38403	.92332	.41592	2.4043	1.0830	2.6040	25
36	.38429	.92321	.41626	.4023	.0832	.6022	24
37	.38456	.92310	.41660	.4004	.0833	.6003	23
38	.38483	.92299	.41694	.3984	.0834	.5985	22
39	.38510	.92287	.41728	.3964	.0836	.5967	21
40	.38537	.92276	.41762	2.3945	1.0837	2.5949	20
41	.38564	.92265	.41797	.3925	.0838	.5931	19
42	.38591	.92254	.41831	.3906	.0840	.5913	18
43	.38617	.92242	.41865	.3886	.0841	.5895	17
44	.38644	.92231	.41899	.3867	.0842	.5877	16
45	.38671	.92220	.41933	2.3847	1.0844	2.5859	15
46	.38698	.92209	.41968	.3828	.0845	.5841	14
47	.38725	.92197	.42002	.3808	.0846	.5823	13
48	.38751	.92186	.42036	.3789	.0847	.5805	12
49	.38778	.92175	.42070	.3770	.0849	.5787	11
50	.38805	.92164	.42105	2.3750	1.0850	2.5770	10
51	.38832	.92152	.42139	.3731	.0851	.5752	9
52	.38859	.92141	.42173	.3712	.0853	.5734	8
53	.38886	.92130	.42207	.3692	.0854	.5716	7
54	.38912	.92118	.42242	.3673	.0855	.5699	6
55	.38939	.92107	.42276	2.3654	1.0857	2.5681	5
56	.38966	.92096	.42310	.3635	.0858	.5663	4
57	.38993	.92084	.42344	.3616	.0859	.5646	3
58	.39019	.92073	.42379	.3597	.0861	.5628	2
59	.39046	.92062	.42413	.3577	.0862	.5610	1
60	.39073	.92050	.42447	2.3558	1.0864	2.5593	0
M	Cosine	Sine	Cotan.	Tan.	Cosec.	Secant	M

67°

51

23°

M	Sine	Cosine	Tan.	Cotsn.	Secant	Cosec.	M
0	.39073	.92050	.42447	2.3558	1.0864	2.5593	60
1	.39100	.92039	.42482	.3539	.0865	.5575	59
2	.39126	.92028	.42516	.3520	.0866	.5558	58
3	.39153	.92016	.42550	.3501	.0868	.5540	57
4	.39180	.92005	.42585	.3482	.0869	.5523	56
5	.39207	.91993	.42619	2.3463	1.0870	2.5506	55
6	.39234	.91982	.42654	.3445	.0872	.5488	54
7	.39260	.91971	.42688	.3426	.0873	.5471	53
8	.39287	.91959	.42722	.3407	.0874	.5453	52
9	.39314	.91948	.42757	.3388	.0876	.5436	51
10	.39341	.91936	.42791	2.3369	1.0877	2.5419	50
11	.39367	.91925	.42826	.3350	.0878	.5402	49
12	.39394	.91913	.428E0	.3332	.0880	.5384	48
13	.39421	.91902	.42894	.3313	.0881	.5367	47
14	.39448	.91891	.42929	.3294	.0882	.5350	46
15	.39474	.91879	.42963	2.3276	1.0884	2.5333	45
16	.39501	.91868	.42998	.3257	.0885	.5316	44
17	.39528	.91856	.43032	.3238	.0886	.5299	43
18	.39554	.91845	.43067	.3220	.0888	.5281	42
19	.39581	.91833	.43101	.3201	.0889	.5264	41
20	.39608	.91822	.43136	2.3183	1.0891	2.5247	40
21	.39635	.91810	.43170	.3164	.0892	.5230	39
22	.39661	.91798	.43205	.3145	.0893	.5213	38
23	.39688	.91787	.43239	.3127	.0895	.5196	37
24	.39715	.91775	.43274	.3109	.0896	.5179	36
25	.39741	.91764	.43308	2.3090	1.0897	2.5163	35
26	.39768	.91752	.43343	.3072	.0899	.5146	34
27	.39795	.91741	.43377	.3053	.0900	.5129	33
28	.39821	.91729	.43412	.3035	.0902	.5112	32
29	.39848	.91718	.43447	.3017	.0903	.5095	31
30	.39875	.91706	.43481	2.2998	1.0904	2.5078	30
31	.39901	.91694	.43516	.2980	.0906	.5062	29
32	.39928	.91683	.43550	.2962	.0907	.5045	28
33	.39955	.91671	.43585	.2944	.0908	.5028	27
34	.39981	.91659	.43620	.2925	.0910	.5011	26
35	.40008	.91648	.43654	2.2907	1.0911	2.4995	25
36	.40035	.91636	.43689	.2889	.0913	.4978	24
37	.40061	.91625	.43723	.2871	.0914	.4961	23
38	.40088	.91613	.43758	.2853	.0915	.4945	22
39	.40115	.91601	.43793	.2835	.0917	.4928	21
40	.40141	.91590	.43827	2.2817	1.0918	2.4912	20
41	.40168	.91578	.43862	.2799	.0920	.4895	19
42	.40195	.91566	.43897	.2781	.0921	.4879	18
43	.40221	.91554	.43932	.2763	.0922	.4862	17
44	.40248	.91543	.43966	.2745	.0924	.4846	16
45	.40275	.91531	.44001	2.2727	1.0925	2.4829	15
46	.40301	.91519	.44036	.2709	.0927	.4813	14
47	.40328	.91508	.44070	.2691	.0928	.4797	13
48	.40354	.91496	.44105	.2673	.0929	.4780	12
49	.40381	.91484	.44140	.2655	.0931	.4764	11
50	.40408	.91472	.44175	2.2637	1.0932	2.4748	10
51	.40434	.91461	.44209	.2619	.0934	.4731	9
52	.40461	.91449	.44244	.2602	.0935	.4715	8
53	.40487	.91437	.44279	.2584	.0936	.4699	7
54	.40514	.91425	.44314	.2566	.0938	.4683	6
55	.40541	.91414	.44349	2.2548	1.0939	2.4666	5
56	.40567	.91402	.44383	.2531	.0941	.4650	4
57	.40594	.91390	.44418	.2513	.0942	.4634	3
58	.40620	.91378	.44453	.2495	.0943	.4618	2
59	.40647	.91366	.44488	.2478	.0945	.4602	1
60	.40674	.91354	.44523	2.2460	1.0946	2.4586	0

M	Cosine	Sine	Cotan.	Tan.	Cosec.	Secant	M

66°

24°

M	Sine	Cosine	Tan.	Cotan.	Secant	Cosec.	M
0	.40674	.91354	.44523	2.2460	1.0946	2.4586	60
1	.40700	.91343	.44558	.2443	.0948	.4570	59
2	.40727	.91331	.44593	.2425	.0949	.4554	58
3	.40753	.91319	.44627	.2408	.0951	.4538	57
4	.40780	.91307	.44662	.2390	.0952	.4522	56
5	.40806	.91295	.44697	2.2373	1.0953	2.4506	55
6	.40833	.91283	.44732	.2355	.0955	.4490	54
7	.40860	.91271	.44767	.2338	.0956	.4474	53
8	.40886	.91260	.44802	.2320	.0958	.4458	52
9	.40913	.91248	.44837	.2303	.0959	.4442	51
10	.40939	.91236	.44872	2.2286	1.0961	2.4426	50
11	.40966	.91224	.44907	.2268	.0962	.4411	49
12	.40992	.91212	.44942	.2251	.0963	.4395	48
13	.41019	.91200	.44977	.2234	.0965	.4379	47
14	.41045	.91188	.45012	.2216	.0966	.4363	46
15	.41072	.91176	.45047	2.2199	1.0968	2.4347	45
16	.41098	.91164	.45082	.2182	.0969	.4332	44
17	.41125	.91152	.45117	.2165	.0971	.4316	43
18	.41151	.91140	.45152	.2147	.0972	.4300	42
19	.41178	.91128	.45187	.2130	.0973	.4285	41
20	.41204	.91116	.45222	2.2113	1.0975	2.4269	40
21	.41231	.91104	.45257	.2096	.0976	.4254	39
22	.41257	.91092	.45292	.2079	.0978	.4238	38
23	.41284	.91080	.45327	.2062	.0979	.4222	37
24	.41310	.91068	.45362	.2045	.0981	.4207	36
25	.41337	.91056	.45397	2.2028	1.0982	2.4191	35
26	.41363	.91044	.45432	.2011	.0984	.4176	34
27	.41390	.91032	.45467	.1994	.0985	.4160	33
28	.41416	.91020	.45502	.1977	.0986	.4145	32
29	.41443	.91008	.45537	.1960	.0988	.4130	31
30	.41469	.90996	.45573	2.1943	1.0989	2.4114	30
31	.41496	.90984	.45608	.1926	.0991	.4099	29
32	.41522	.90972	.45643	.1909	.0992	.4083	28
33	.41549	.90960	.45678	.1892	.0994	.4068	27
34	.41575	.90948	.45713	.1875	.0995	.4053	26
35	.41602	.90936	.45748	2.1859	1.0997	2.4037	25
36	.41628	.90924	.45783	.1842	.0998	.4022	24
37	.41654	.90911	.45819	.1825	.1000	.4007	23
38	.41681	.90899	.45854	.1808	.1001	.3992	22
39	.41707	.90887	.45889	.1792	.1003	.3976	21
40	.41734	.90875	.45924	2.1775	1.1004	2.3961	20
41	.41760	.90863	.45960	.1758	.1005	.3946	19
42	.41787	.90851	.45995	.1741	.1007	.3931	18
43	.41813	.90839	.46030	.1725	.1008	.3916	17
44	.41839	.90826	.46065	.1708	.1010	.3901	16
45	.41866	.90814	.46101	2.1692	1.1011	2.3886	15
46	.41892	.90802	.46136	.1675	.1013	.3871	14
47	.41919	.90790	.46171	.1658	.1014	.3856	13
48	.41945	.90778	.46206	.1642	.1016	.3841	12
49	.41972	.90765	.46242	.1625	.1017	.3826	11
50	.41998	.90753	.46277	2.1609	1.1019	2.3811	10
51	.42024	.90741	.46312	.1592	.1020	.3796	9
52	.42051	.90729	.46348	.1576	.1022	.3781	8
53	.42077	.90717	.46383	.1559	.1023	.3766	7
54	.42103	.90704	.46418	.1543	.1025	.3751	6
55	.42130	.90692	.46454	2.1527	1.1026	2.3736	5
56	.42156	.90680	.46489	.1510	.1028	.3721	4
57	.42183	.90668	.46524	.1494	.1029	.3706	3
58	.42209	.90655	.46560	.1478	.1031	.3691	2
59	.42235	.90643	.46595	.1461	.1032	.3677	1
60	.42262	.90631	.46631	.1445	1.1034	2.3662	0

M	Cosine	Sine	Cotan.	Tan.	Cosec.	Secant	M

65°

25°

M	Sine	Cosine	Tan.	Cotan.	Secant	Cosec.	M
0	.42262	.90631	.46631	2.1445	1.1034	2.3662	60
1	.42288	.90618	.46666	.1429	.1035	.3647	59
2	.42314	.90606	.46702	.1412	.1037	.3632	58
3	.42341	.90594	.46737	.1396	.1038	.3618	57
4	.42367	.90581	.46772	.1380	.1040	.3603	56
5	.42394	.90569	.46808	2.1364	1.1041	2.3588	55
6	.42420	.90557	.46843	.1348	.1043	.3574	54
7	.42446	.90545	.46879	.1331	.1044	.3559	53
8	.42473	.90532	.46914	.1315	.1046	.3544	52
9	.42499	.90520	.46950	.1299	.1047	.3530	51
10	.42525	.90507	.46985	2.1283	1.1049	2.3515	50
11	.42552	.90495	.47021	.1267	.1050	.3501	49
12	.42578	.90483	.47056	.1251	.1052	.3486	48
13	.42604	.90470	.47092	.1235	.1053	.3472	47
14	.42630	.90458	.47127	.1219	.1055	.3457	46
15	.42657	.90445	.47163	2.1203	1.1056	2.3443	45
16	.42683	.90433	.47199	.1187	.1058	.3428	44
17	.42709	.90421	.47234	.1171	.1059	.3414	43
18	.42736	.90408	.47270	.1155	.1061	.3399	42
19	.42762	.90396	.47305	.1139	.1062	.3385	41
20	.42788	.90383	.47341	2.1123	1.1064	2.3371	40
21	.42815	.90371	.47376	.1107	.1065	.3356	39
22	.42841	.90358	.47412	.1092	.1067	.3342	38
23	.42867	.90346	.47448	.1076	.1068	.3328	37
24	.42893	.90333	.47483	.1060	.1070	.3313	36
25	.42920	.90321	.47519	2.1044	1.1072	2.3299	35
26	.42946	.90308	.47555	.1028	.1073	.3285	34
27	.42972	.90296	.47590	.1013	.1075	.3271	33
28	.42998	.90283	.47626	.0997	.1076	.3256	32
29	.43025	.90271	.47662	.0981	.1078	.3242	31
30	.43051	.90258	.47697	2.0965	1.1079	2.3228	30
31	.43077	.90246	.47733	.0950	.1081	.3214	29
32	.43104	.90233	.47769	.0934	.1082	.3200	28
33	.43130	.90221	.47805	.0918	.1084	.3186	27
34	.43156	.90208	.47840	.0903	.1085	.3172	26
35	.43182	.90196	.47876	2.0887	1.1087	2.3158	25
36	.43208	.90183	.47912	.0872	.1088	.3143	24
37	.43235	.90171	.47948	.0856	.1090	.3129	23
38	.43261	.90158	.47983	.0840	.1092	.3115	22
39	.43287	.90145	.48019	.0825	.1093	.3101	21
40	.43313	.90133	.48055	2.0809	1.1095	2.3087	20
41	.43340	.90120	.48091	.0794	.1096	.3073	19
42	.43366	.90108	.48127	.0778	.1098	.3059	18
43	.43392	.90095	.48162	.0763	.1099	.3046	17
44	.43418	.90082	.48198	.0747	.1101	.3032	16
45	.43444	.90070	.48234	2.0732	1.1102	2.3018	15
46	.43471	.90057	.48270	.0717	.1104	.3004	14
47	.43497	.90044	.48306	.0701	.1106	.2990	13
48	.43523	.90032	.48342	.0686	.1107	.2976	12
49	.43549	.90019	.48378	.0671	.1109	.2962	11
50	.43575	.90006	.48414	2.0655	1.1110	2.2949	10
51	.43602	.89994	.48449	.0640	.1112	.2935	9
52	.43628	.89981	.48485	.0625	.1113	.2921	8
53	.43654	.89968	.48521	.0609	.1115	.2907	7
54	.43680	.89956	.48557	.0594	.1116	.2894	6
55	.43706	.89943	.48593	2.0579	1.1118	2.2880	5
56	.43732	.89930	.48629	.0564	.1120	.2866	4
57	.43759	.89918	.48665	.0548	.1121	.2853	3
58	.43785	.89905	.48701	.0533	.1123	.2839	2
59	.43811	.89892	.48737	.0518	.1124	.2825	1
60	.43837	.89879	.48773	.0503	.1126	.2812	0
M	Cosine	Sine	Cotan.	Tan.	Cosec.	Secant	M

64°

26°

M	Sine	Cosine	Tan.	Cotan.	Secant	Cosec.	M
0	.43837	.89879	.48773	2.0503	1.1126	2.2812	60
1	.43863	.89867	.48809	.0488	.1127	.2798	59
2	.43889	.89854	.48845	.0473	.1129	.2784	58
3	.43915	.89841	.48881	.0458	.1131	.2771	57
4	.43942	.89828	.48917	.0443	.1132	.2757	56
5	.43968	.89815	.48953	2.0427	1.1134	2.2744	55
6	.43994	.89803	.48989	.0412	.1135	.2730	54
7	.44020	.89790	.49025	.0397	.1137	.2717	53
8	.44046	.89777	.49062	2.0382	.1139	.2703	52
9	.44072	.89764	.49098	.0367	.1140	.2690	51
10	.44098	.89751	.49134	2.0352	1.1142	2.2676	50
11	.44124	.89739	.49170	.0338	.1143	.2663	49
12	.44150	.89726	.49206	.0323	.1145	.2650	48
13	.44177	.89713	.49242	.0308	.1147	.2636	47
14	.44203	.89700	.49278	.0293	.1148	.2623	46
15	.44229	.89687	.49314	2.0278	1.1150	2.2610	45
16	.44255	.89674	.49351	.0263	.1151	.2596	44
17	.44281	.89661	.49387	.0248	.1153	.2583	43
18	.44307	.89649	.49423	.0233	.1155	.2570	42
19	.44333	.89636	.49459	.0219	.1156	.2556	41
20	.44359	.89623	.49495	2.0204	1.1158	2.2543	40
21	.44385	.89610	.49532	.0189	.1159	.2530	39
22	.44411	.89597	.49568	.0174	.1161	.2517	38
23	.44437	.89584	.49604	.0159	.1163	.2503	37
24	.44463	.89571	.49640	.0145	.1164	.2490	36
25	.44489	.89558	.49677	2.0130	1.1166	2.2477	35
26	.44516	.89545	.49713	.0115	.1167	.2464	34
27	.44542	.89532	.49749	.0101	.1169	.2451	33
28	.44568	.89519	.49785	.0086	.1171	.2438	32
29	.44594	.89506	.49822	.0071	.1172	.2425	31
30	.44620	.89493	.49858	2.0057	1.1174	2.2411	30
31	.44646	.89480	.49894	.0042	.1176	.2398	29
32	.44672	.89467	.49931	.0028	.1177	.2385	28
33	.44698	.89454	.49967	.0013	.1179	.2372	27
34	.44724	.89441	.50003	1.9998	.1180	.2359	26
35	.44750	.89428	.50040	1.9984	1.1182	2.2348	25
36	.44776	.89415	.50076	.9969	.1184	.2333	24
37	.44802	.89402	.50113	.9955	.1185	.2320	23
38	.44828	.89389	.50149	.9940	.1187	.2307	22
39	.44854	.89376	.50185	.9926	.1189	.2294	21
40	.44880	.89363	.50222	1.9912	1.1190	2.2282	20
41	.44906	.89350	.50258	.9897	.1192	.2269	19
42	.44932	.89337	.50295	.9883	.1193	.2256	18
43	.44958	.89324	.50331	.9868	.1195	.2243	17
44	.44984	.89311	.50368	.9854	.1197	.2230	16
45	.45010	.89298	.50404	1.9840	1.1198	2.2217	15
46	.45036	.89285	.50441	.9825	.1200	.2204	14
47	.45062	.89272	.50477	.9811	.1202	.2192	13
48	.45088	.89258	.50514	.9797	.1203	.2179	12
49	.45114	.89245	.50550	.9782	.1205	.2166	11
50	.45140	.89232	.50587	1.9768	1.1207	2.2153	10
51	.45166	.89219	.50623	.9754	.1208	.2141	9
52	.45191	.89206	.50660	.9739	.1210	.2128	8
53	.45217	.89193	.50696	.9725	.1212	.2115	7
54	.45243	.89180	.50733	.9711	.1213	.2103	6
55	.45269	.89166	.50769	1.9697	1.1215	2.2090	5
56	.45295	.89153	.50806	.9683	.1217	.2077	4
57	.45321	.89140	.50843	.9668	.1218	.2065	3
58	.45347	.89127	.50879	.9654	.1220	.2052	2
59	.45373	.89114	.50916	.9640	.1222	.2039	1
60	.45399	.89101	.50952	1.9626	1.1223	2.2027	0

| M | Cosine | Sine | Cotan. | Tan. | Cosec. | Secant | M |

63°

27°

M	Sine	Cosine	Tan.	Cotan.	Secant	Cosec.	M
0	.45399	.89101	.50952	1.9626	1.1223	2.2027	60
1	.45425	.89087	.50989	.9612	.1225	.2014	59
2	.45451	.89074	.51026	.9598	.1226	.2002	58
3	.45477	.89061	.51062	.9584	.1228	.1989	57
4	.45503	.89048	.51099	.9570	.1230	.1977	56
5	.45528	.89034	.51136	1.9556	1.1231	2.1964	55
6	.45554	.89021	.51172	.9542	.1233	.1952	54
7	.45580	.89008	.51209	.9528	.1235	.1939	53
8	.45606	.88995	.51246	.9514	.1237	.1927	52
9	.45632	.88981	.51283	.9500	.1238	.1914	51
10	.45658	.88968	.51319	1.9486	1.1240	2.1902	50
11	.45684	.88955	.51356	.9472	.1242	.1889	49
12	.45710	.88942	.51393	.9458	.1243	.1877	48
13	.45736	.88928	.51430	.9444	.1245	.1865	47
14	.45761	.88915	.51466	.9430	.1247	.1852	46
15	.45787	.88902	.51503	1.9416	1.1248	2.1840	45
16	.45813	.88888	.51540	.9402	.1250	.1828	44
17	.45839	.88875	.51577	.9388	.1252	.1815	43
18	.45865	.88862	.51614	.9375	.1253	.1803	42
19	.45891	.88848	.51651	.9361	.1255	.1791	41
20	.45917	.88835	.51687	1.9347	1.1257	2.1778	40
21	.45942	.88822	.51724	.9333	.1258	.1766	39
22	.45968	.88808	.51761	.9319	.1260	.1754	38
23	.45994	.88795	.51798	.9306	.1262	.1742	37
24	.46020	.88781	.51835	.9292	.1264	.1730	36
25	.46046	.88768	.51872	1.9278	1.1265	2.1717	35
26	.46072	.88755	.51909	.9264	.1267	.1705	34
27	.46097	.88741	.51946	.9251	.1269	.1693	33
28	.46123	.88728	.51983	.9237	.1270	.1681	32
29	.46149	.88714	.52020	.9223	.1272	.1669	31
30	.46175	.88701	.52057	1.9210	1.1274	2.1657	30
31	.46201	.88688	.52094	.9196	.1275	.1645	29
32	.46226	.88674	.52131	.9182	.1277	.1633	28
33	.46252	.88661	.52168	.9169	.1279	.1620	27
34	.46278	.88647	.52205	.9155	.1281	.1608	26
35	.46304	.88634	.52242	1.9142	1.1282	2.1596	25
36	.46330	.88620	.52279	.9128	.1284	.1584	24
37	.46355	.88607	.52316	.9115	.1286	.1572	23
38	.46381	.88593	.52353	.9101	.1287	.1560	22
39	.46407	.88580	.52390	.9088	.1289	.1548	21
40	.46433	.88566	.52427	1.9074	1.1291	2.1536	20
41	.46458	.88553	.52464	.9061	.1293	.1525	19
42	.46484	.88539	.52501	.9047	.1294	.1513	18
43	.46510	.88526	.52538	.9034	.1296	.1501	17
44	.46536	.88512	.52575	.9020	.1298	.1489	16
45	.46561	.88499	.52612	1.9007	1.1299	2.1477	15
46	.46587	.88485	.52650	.8993	.1301	.1465	14
47	.46613	.88472	.52687	.8980	.1303	.1453	13
48	.46639	.88458	.52724	.8967	.1305	.1441	12
49	.46664	.88444	.52761	.8953	.1306	.1430	11
50	.46690	.88431	.52798	1.8940	1.1308	2.1418	10
51	.46716	.88417	.52836	.8927	.1310	.1406	9
52	.46741	.88404	.52873	.8913	.1312	.1394	8
53	.46767	.88390	.52910	.8900	.1313	.1382	7
54	.46793	.88376	.52947	.8887	.1315	.1371	6
55	.46819	.88363	.52984	1.8873	1.1317	2.1359	5
56	.46844	.88349	.53022	.8860	.1319	.1347	4
57	.46870	.88336	.53059	.8847	.1320	.1335	3
58	.46896	.88322	.53096	.8834	.1322	.1324	2
59	.46921	.88308	.53134	.8820	.1324	.1312	1
60	.46947	.88295	.53171	1.8807	1.1326	2.1300	0
M	Cosine	Sine	Cotan.	Tan.	Cosec.	Secant	M

62°

28°

M	Sine	Cosine	Tan.	Cotan.	Secant	Cosec.	M
0	.46947	.88295	.53171	1.8807	1.1326	2.1300	60
1	.46973	.88281	.53208	.8794	.1327	.1289	59
2	.46998	.88267	.53245	.8781	.1329	.1277	58
3	.47024	.88254	.53283	.8768	.1331	.1266	57
4	.47050	.88240	.53320	.8754	.1333	.1254	56
5	.47075	.88226	.53358	1.8741	1.1334	2.1242	55
6	.47101	.88213	.53395	.8728	.1336	.1231	54
7	.47127	.88199	.53432	.8715	.1338	.1219	53
8	.47152	.88185	.53470	.8702	.1340	.1208	52
9	.47178	.88171	.53507	.8689	.1341	.1196	51
10	.47204	.88158	.53545	1.8676	1.1343	2.1185	50
11	.47229	.88144	.53582	.8663	.1345	.1173	49
12	.47255	.88130	.53619	.8650	.1347	.1162	48
13	.47281	.88117	.53657	.8637	.1349	.1150	47
14	.47306	.88103	.53694	.8624	.1350	.1139	46
15	.47332	.88089	.53732	1.8611	1.1352	2.1127	45
16	.47357	.88075	.53769	.8598	.1354	.1116	44
17	.47383	.88061	.53807	.8572	.1356	.1104	43
18	.47409	.88048	.53844	.8572	.1357	.1093	42
19	.47434	.88034	.53882	.8559	.1359	.1082	41
20	.47460	.88020	.53919	1.8546	1.1361	2.1070	40
21	.47486	.88006	.53957	.8533	.1363	.1059	39
22	.47511	.87992	.53995	.8520	.1365	.1048	38
23	.47537	.87979	.54032	.8507	.1366	.1036	37
24	.47562	.87965	.54070	.8495	.1368	.1025	36
25	.47588	.87951	.54107	1.8482	1.1370	2.1014	35
26	.47613	.87937	.54145	.8469	.1372	.1002	34
27	.47639	.87923	.54183	.8456	.1373	.0991	33
28	.47665	.87909	.54220	.8443	.1375	.0980	32
29	.47690	.87895	.54258	.8430	.1377	.0969	31
30	.47716	.87882	.54295	1.8418	1.1379	2.0957	30
31	.47741	.87868	.54333	.8405	.1381	.0946	29
32	.47767	.87854	.54371	.8392	.1382	.0935	28
33	.47792	.87840	.54409	.8379	.1384	.0924	27
34	.47818	.87826	.54446	.8367	.1386	.0912	26
35	.47844	.87812	.54484	1.8354	1.1388	2.0901	25
36	.47869	.87798	.54522	.8341	.1390	.0890	24
37	.47895	.87784	.54559	.8329	.1391	.0879	23
38	.47920	.87770	.54597	.8316	.1393	.0868	22
39	.47946	.87756	.54635	.8303	.1395	.0857	21
40	.47971	.87742	.54673	1.8291	1.1397	2.0846	20
41	.47997	.87728	.54711	.8278	.1399	.0835	19
42	.48022	.87715	.54748	.8265	.1401	.0824	18
43	.48048	.87701	.54786	.8253	.1402	.0812	17
44	.48073	.87687	.54824	.8240	.1404	.0801	16
45	.48099	.87673	.54862	1.8227	1.1406	2.0790	15
46	.48124	.87659	.54900	.8215	.1408	.0779	14
47	.48150	.87645	.54937	.8202	.1410	.0768	13
48	.48175	.87631	.54975	.8190	.1411	.0757	12
49	.48201	.87617	.55013	.8177	.1413	.0746	11
50	.48226	.87603	.55051	1.8165	1.1415	2.0735	10
51	.48252	.87588	.55089	.8152	.1417	.0725	9
52	.48277	.87574	.55127	.8140	.1419	.0714	8
53	.48303	.87560	.55165	.8127	.1421	.0703	7
54	.48328	.87546	.55203	.8115	.1422	.0692	6
55	.48354	.87532	.55241	1.8102	1.1424	2.0681	5
56	.48379	.87518	.55279	.8090	.1426	.0670	4
57	.48405	.87504	.55317	.8078	.1428	.0659	3
58	.48430	.87490	.55355	.8065	.1430	.0648	2
59	.48455	.87476	.55393	.8053	.1432	.0637	1
60	.48481	.87462	.55431	1.8040	1.1433	2.0627	0
M	Cosine	Sine	Cotan.	Tan.	Cosec.	Secant	M

61°

29°

M	Sine	Cosine	Tan.	Cotan.	Secant	Cosec.	M
0	.48481	.87462	.55431	1.8040	1.1433	2.0627	60
1	.48506	.87448	.55469	.8028	.1435	.0616	59
2	.48532	.87434	.55507	.8016	.1437	.0605	58
3	.48557	.87420	.55545	.8003	.1439	.0594	57
4	.48583	.87405	.55583	.7991	.1441	.0583	56
5	.48608	.87391	.55621	1.7979	1.1443	2.0573	55
6	.48633	.87377	.55659	.7966	.1445	.0562	54
7	.48659	.87363	.55697	.7954	.1446	.0551	53
8	.48684	.87349	.55735	.7942	.1448	.0540	52
9	.48710	.87335	.55774	.7930	.1450	.0530	51
10	.48735	.87320	.55812	1.7917	1.1452	2.0519	50
11	.48760	.87306	.55850	.7905	.1454	.0508	49
12	.48786	.87292	.55888	.7893	.1456	.0498	48
13	.48811	.87278	.55926	.7881	.1458	.0487	47
14	.48837	.87264	.55964	.7868	.1459	.0476	46
15	.48862	.87250	.56003	1.7856	1.1461	2.0466	45
16	.48887	.87235	.56041	.7844	.1463	.0455	44
17	.48913	.87221	.56079	.7832	.1465	.0444	43
18	.48938	.87207	.56117	.7820	.1467	.0434	42
19	.48964	.87193	.56156	.7808	.1469	.0423	41
20	.48989	.87178	.56194	1.7795	1.1471	2.0413	40
21	.49014	.87164	.56232	.7783	.1473	.0402	39
22	.49040	.87150	.56270	.7771	.1474	.0392	38
23	.49065	.87136	.56309	.7759	.1476	.0381	37
24	.49090	.87121	.56347	.7747	.1478	.0370	36
25	.49116	.87107	.56385	1.7735	1.1480	2.0360	35
26	.49141	.87093	.56424	.7723	.1482	.0349	34
27	.49166	.87078	.56462	.7711	.1484	.0339	33
28	.49192	.87064	.56500	.7699	.1486	.0329	32
29	.49217	.87050	.56539	.7687	.1488	.0318	31
30	.49242	.87035	.56577	1.7675	1.1489	2.0308	30
31	.49268	.87021	.56616	.7663	.1491	.0297	29
32	.49293	.87007	.56654	.7651	.1493	.0287	28
33	.49318	.86992	.56692	.7639	.1495	.0276	27
34	.49343	.86978	.56731	.7627	.1497	.0266	26
35	.49369	.86964	.56769	1.7615	1.1499	2.0256	25
36	.49394	.86949	.56808	.7603	.1501	.0245	24
37	.49419	.86935	.56846	.7591	.1503	.0235	23
38	.49445	.86921	.56885	.7579	.1505	.0224	22
39	.49470	.86906	.56923	.7567	.1507	.0214	21
40	.49495	.86892	.56962	1.7555	1.1508	2.0204	20
41	.49521	.86877	.57000	.7544	.1510	.0194	19
42	.49546	.86863	.57039	.7532	.1512	.0183	18
43	.49571	.86849	.57077	.7520	.1514	.0173	17
44	.49596	.86834	.57116	.7508	.1516	.0163	16
45	.49622	.86820	.57155	1.7496	1.1518	2.0152	15
46	.49647	.86805	.57193	.7484	.1520	.0142	14
47	.49672	.86791	.57232	.7473	.1522	.0132	13
48	.49697	.86776	.57270	.7461	.1524	.0122	12
49	.49723	.86762	.57309	.7449	.1526	.0111	11
50	.49748	.86748	.57348	1.7437	1.1528	2.0101	10
51	.49773	.86733	.57386	.7426	.1530	.0091	9
52	.49798	.86719	.57425	.7414	.1531	.0081	8
53	.49823	.86704	.57464	.7402	.1533	.0071	7
54	.49849	.86690	.57502	.7390	.1535	.0061	6
55	.49874	.86675	.57541	1.7379	1.1537	2.0050	5
56	.49899	.86661	.57580	.7367	.1539	.0040	4
57	.49924	.86646	.57619	.7355	.1541	.0030	3
58	.49950	.86632	.57657	.7344	.1543	.0020	2
59	.49975	.86617	.57696	.7332	.1545	.0010	1
60	.50000	.86603	.57735	1.7320	1.1547	2.0000	0

M	Cosine	Sine	Cotan.	Tan.	Cosec.	Secant	M

60°

30°

M	Sine	Cosine	Tan.	Cotan.	Secant	Cosec.	M
0	.50000	.86603	.57735	1.7320	1.1547	2.0000	60
1	.50025	.86588	.57774	.7309	.1549	1.9990	59
2	.50050	.86573	.57813	.7297	.1551	.9980	58
3	.50075	.86559	.57851	.7286	.1553	.9970	57
4	.50101	.86544	.57890	.7274	.1555	.9960	56
5	.50126	.86530	.57929	1.7262	1.1557	1.9950	55
6	.50151	.86515	.57968	.7251	.1559	.9940	54
7	.50176	.86500	.58007	.7239	.1561	.9930	53
8	.50201	.86486	.58046	.7228	.1562	.9920	52
9	.50226	.86471	.58085	.7216	.1564	.9910	51
10	.50252	.86457	.58123	1.7205	1.1566	1.9900	50
11	.50277	.86442	.58162	.7193	.1568	.9890	49
12	.50302	.86427	.58201	.7182	.1570	.9880	48
13	.50327	.86413	.58240	.7170	.1572	.9870	47
14	.50352	.86398	.58279	.7159	.1574	.9860	46
15	.50377	.86383	.58318	1.7147	1.1576	1.9850	45
16	.50402	.86369	.58357	.7136	.1578	.9840	44
17	.50428	.86354	.58396	.7124	.1580	.9830	43
18	.50453	.86339	.58435	.7113	.1582	.9820	42
19	.50478	.86325	.58474	.7101	.1584	.9811	41
20	.50503	.86310	.58513	1.7090	1.1586	1.9801	40
21	.50528	.86295	.58552	.7079	.1588	.9791	39
22	.50553	.86281	.58591	.7067	.1590	.9781	38
23	.50578	.86266	.58630	.7056	.1592	.9771	37
24	.50603	.86251	.58670	.7044	.1594	.9761	36
25	.50628	.86237	.58709	1.7033	1.1596	1.9752	35
26	.50653	.86222	.58748	.7022	.1598	.9742	34
27	.50679	.86207	.58787	.7010	.1600	.9732	33
28	.50704	.86192	.58826	.6999	.1602	.9722	32
29	.50729	.86178	.58865	.6988	.1604	.9713	31
30	.50754	.86163	.58904	1.6977	1.1606	1.9703	30
31	.50779	.86148	.58944	.6965	.1608	.9693	29
32	.50804	.86133	.58983	.6954	.1610	.9683	28
33	.50829	.86118	.59022	.6943	.1612	.9674	27
34	.50854	.86104	.59061	.6931	.1614	.9664	26
35	.50879	.86089	.59100	1.6920	1.1616	1.9654	25
36	.50904	.86074	.59140	.6909	.1618	.9645	24
37	.50929	.86059	.59179	.6898	.1620	.9635	23
38	.50954	.86044	.59218	.6887	.1622	.9625	22
39	.50979	.86030	.59258	.6875	.1624	.9616	21
40	.51004	.86015	.59297	1.6864	1.1626	1.9606	20
41	.51029	.86000	.59336	.6853	.1628	.9596	19
42	.51054	.85985	.59376	.6842	.1630	9587	18
43	.51079	.85970	.59415	.6831	.1632	.9577	17
44	.51104	.85955	.59454	.6820	.1634	.9568	16
45	.51129	.85941	.59494	1.6808	1.1636	1.9558	15
46	.51154	.85926	.59533	.6797	.1638	.9549	14
47	.51179	.85911	.59572	.6786	.1640	.9539	13
48	.51204	.85896	.59612	.6775	.1642	.9530	12
49	.51229	.85881	.59651	.6764	.1644	.9520	11
50	.51254	.85866	.59691	1.6753	1.1646	1.9510	10
51	.51279	.85851	.59730	.6742	.1648	.9501	9
52	.51304	.85836	.59770	.6731	.1650	.9491	8
53	.51329	.85821	.59809	.6720	.1652	.9482	7
54	.51354	.85806	.59849	.6709	.1654	.9473	6
55	.51379	.85791	.59888	1.6698	1.1656	1.9463	5
56	.51404	.85777	.59928	.6687	.1658	.9454	4
57	.51429	.85762	.59967	.6676	.1660	.9444	3
58	.51454	.85747	.60007	.6665	.1662	.9435	2
59	.51479	.85732	.60046	.6654	.1664	.9425	1
60	.51504	.85717	.60086	1.6643	1.1666	1.9416	0

M	Cosine	Sine	Cotan.	Tan.	Cosec.	Secant	M

59°

31°

M	Sine	Cosine	Tan.	Cotan.	Secant	Cosec.	M
0	.51504	.85717	.60086	1.6643	1.1666	1.9416	60
1	.51529	.85702	.60126	.6632	.1668	.9407	59
2	.51554	.85687	.60165	.6621	.1670	.9397	58
3	.51578	.85672	.60205	.6610	.1672	.9388	57
4	.51603	.85657	.60244	.6599	.1674	.9378	56
5	.51628	.85642	.60284	1.6588	1.1676	1.9369	55
6	.51653	.85627	.60324	.6577	.1678	.9360	54
7	.51678	.85612	.60363	.6566	.1681	.9350	53
8	.51703	.85597	.60403	.6555	.1683	.9341	52
9	.51728	.85582	.60443	.6544	.1685	.9332	51
10	.51753	.85566	.60483	1.6534	1.1687	1.9322	50
11	.51778	.85551	.60522	.6523	.1689	.9313	49
12	.51803	.85536	.60562	.6512	.1691	.9304	48
13	.51827	.85521	.60602	.6501	.1693	.9295	47
14	.51852	.85506	.60642	.6490	.1695	.9285	46
15	.51877	.85491	.60681	1.6479	1.1697	1.9276	45
16	.51902	.85476	.60721	.6469	.1699	.9267	44
17	.51927	.85461	.60761	.6458	.1701	.9258	43
18	.51952	.85446	.60801	.6447	.1703	.9248	42
19	.51977	.85431	.60841	.6436	.1705	.9239	41
20	.52002	.85416	.60881	1.6425	1.1707	1.9230	40
21	.52026	.85400	.60920	.6415	.1709	.9221	39
22	.52051	.85385	.60960	.6404	.1712	.9212	38
23	.52076	.85370	.61000	.6393	.1714	.9203	37
24	.52101	.85355	.61040	.6383	.1716	.9193	36
25	.52126	.85340	.61080	1.6372	1.1718	1.9184	35
26	.52151	.85325	.61120	.6361	.1720	.9175	34
27	.52175	.85309	.61160	.6350	.1722	.9166	33
28	.52200	.85294	.61200	.6340	.1724	.9157	32
29	.52225	.85279	.61240	.6329	.1726	.9148	31
30	.52250	.85264	.61280	1.6318	1.1728	1.9139	30
31	.52275	.85249	.61320	.6308	.1730	.9130	29
32	.52299	.85234	.61360	.6297	.1732	.9121	28
33	.52324	.85218	.61400	.6286	.1734	.9112	27
34	.52349	.85203	.61440	.6276	.1737	.9102	26
35	.52374	.85188	.61480	1.6265	1.1739	1.9093	25
36	.52398	.85173	.61520	.6255	.1741	.9084	24
37	.52423	.85157	.61560	.6244	.1743	.9075	23
38	.52448	.85142	.61601	.6233	.1745	.9066	22
39	.52473	.85127	.61641	.6223	.1747	.9057	21
40	.52498	.85112	.61681	1.6212	1.1749	1.9048	20
41	.52522	.85096	.61721	.6202	.1751	.9039	19
42	.52547	.85081	.61761	.6191	.1753	.9030	18
43	.52572	.85066	.61801	.6181	.1756	.9021	17
44	.52597	.85050	.61842	.6170	.1758	.9013	16
45	.52621	.85035	.61882	1.6160	1.1760	1.9004	15
46	.52646	.85020	.61922	.6149	.1762	.8995	14
47	.52671	.85004	.61962	.6139	.1764	.8986	13
48	.52695	.84989	.62003	.6128	.1766	.8977	12
49	.52720	.84974	.62043	.6118	.1768	.8968	11
50	.52745	.84959	.62083	1.6107	1.1770	1.8959	10
51	.52770	.84943	.62123	.6097	.1772	.8950	9
52	.52794	.84928	.62164	.6086	.1775	.8941	8
53	.52819	.84912	.62204	.6076	.1777	.8932	7
54	.52844	.84897	.62244	.6066	.1779	.8924	6
55	.52868	.84882	.62285	1.6055	1.1781	1.8915	5
56	.52893	.84866	.62325	.6045	.1783	.8906	4
57	.52918	.84851	.62366	.6034	.1785	.8897	3
58	.52942	.84836	.62406	.6024	.1787	.8888	2
59	.52967	.84820	.62446	.6014	.1790	.8879	1
60	.52992	.84805	.62487	1.6003	1.1792	1.8871	0

M	Cosine	Sine	Cotan.	Tan.	Cosec.	Secant	M

58°

32°

M	Sine	Cosine	Tan.	Cotan.	Secant	Cosec.	M
0	.52992	.84805	.62487	1.6003	1.1792	1.8871	60
1	.53016	.84789	.62527	.5993	.1794	.8862	59
2	.53041	.84774	.62568	.5983	.1796	.8853	58
3	.53066	.84758	.62608	.5972	.1798	.8844	57
4	.53090	.84743	.62649	.5962	.1800	.8836	56
5	.53115	.84728	.62689	1.5952	1.1802	1.8827	55
6	.53140	.84712	.62730	.5941	.1805	.8818	54
7	.53164	.84697	.62770	.5931	.1807	.8809	53
8	.53189	.84681	.62811	.5921	.1809	.8801	52
9	.53214	.84666	.62851	.5910	.1811	.8792	51
10	.53238	.84650	.62892	1.5900	1.1813	1.8783	50
11	.53263	.84635	.62933	.5890	.1815	.8775	49
12	.53288	.84619	.62973	.5880	.1818	.8766	48
13	.53312	.84604	.63014	.5869	.1820	.8757	47
14	.53337	.84588	.63055	.5859	.1822	.8749	46
15	.53361	.84573	.63095	1.5849	1.1824	1.8740	45
16	.53386	.84557	.63136	.5839	.1826	.8731	44
17	.53411	.84542	.63177	.5829	.1828	.8723	43
18	.53435	.84526	.63217	.5818	.1831	.8714	42
19	.53460	.84511	.63258	.5808	.1833	.8706	41
20	.53484	.84495	.63299	1.5798	1.1835	1.8697	40
21	.53509	.84479	.63339	.5788	.1837	.8688	39
22	.53533	.84464	.63380	.5778	.1839	.8680	38
23	.53558	.84448	.63421	.5768	.1841	.8671	37
24	.53583	.84433	.63462	.5757	.1844	.8663	36
25	.53607	.84417	.63503	1.5747	1.1846	1.8654	35
26	.53632	.84402	.63543	.5737	.1848	.8646	34
27	.53656	.84386	.63584	.5727	.1850	.8637	33
28	.53681	.84370	.63625	.5717	.1852	.8629	32
29	.53705	.84355	.63666	.5707	.1855	.8620	31
30	.53730	.84339	.63707	1.5697	1.1857	1.8611	30
31	.53754	.84323	.63748	.5687	.1859	.8603	29
32	.53779	.84308	.63789	.5677	.1861	.8595	28
33	.53803	.84292	.63830	.5667	.1863	.8586	27
34	.53828	.84276	.63871	.5657	.1866	.8578	26
35	.53852	.84261	.63912	1.5646	1.1868	1.8569	25
36	.53877	.84245	.63953	.5636	.1870	.8561	24
37	.53901	.84229	.63994	.5626	.1872	.8552	23
38	.53926	.84214	.64035	.5616	.1874	.8544	22
39	.53950	.84198	.64076	.5606	.1877	.8535	21
40	.53975	.84182	.64117	1.5596	1.1879	1.8527	20
41	.53999	.84167	.64158	.5586	.1881	.8519	19
42	.54024	.84151	.64199	.5577	.1883	.8510	18
43	.54048	.84135	.64240	.5567	.1886	.8502	17
44	.54073	.84120	.64281	.5557	.1888	.8493	16
45	.54097	.84104	.64322	1.5547	1.1890	1.8485	15
46	.54122	.84088	.64363	.5537	.1892	.8477	14
47	.54146	.84072	.64404	.5527	.1894	.8468	13
48	.54171	.84057	.64446	.5517	.1897	.8460	12
49	.54195	.84041	.64487	.5507	.1899	.8452	11
50	.54220	.84025	.64528	1.5497	1.1901	1.8443	10
51	.54244	.84009	.64569	.5487	.1903	.8435	9
52	.54268	.83993	.64610	.5477	.1906	.8427	8
53	.54293	.83978	.64652	.5467	.1908	.8418	7
54	.54317	.83962	.64693	.5458	.1910	.8410	6
55	.54342	.83946	.64734	1.5448	1.1912	1.8402	5
56	.54366	.83930	.64775	.5438	.1915	.8394	4
57	.54391	.83914	.64817	.5428	.1917	.8385	3
58	.54415	.83899	.64858	.5418	.1919	.8377	2
59	.54439	.83883	.64899	.5408	.1921	.8369	1
60	.54464	.83867	.64941	1.5399	1.1924	1.8361	0
M	Cosine	Sine	Cotan.	Tan.	Cosec.	Secant	M

57°

61

33°

M	Sine	Cosine	Tan.	Cotan.	Secant	Cosec.	M
0	.54464	.83867	.64941	1.5399	1.1924	1.8361	60
1	.54488	.83851	.64982	.5389	.1926	.8352	59
2	.54513	.83835	.65023	.5379	.1928	.8344	58
3	.54537	.83819	.65065	.5369	.1930	.8336	57
4	.54561	.83804	.65106	.5359	.1933	.8328	56
5	.54586	.83788	.65148	1.5350	1.1935	1.8320	55
6	.54610	.83772	.65189	.5340	.1937	.8311	54
7	.54634	.83756	.65231	.5330	.1939	.8303	53
8	.54659	.83740	.65272	.5320	.1942	.8295	52
9	.54683	.83724	.65314	.5311	.1944	.8287	51
10	.54708	.83708	.65355	1.5301	1 1946	1.8279	50
11	.54732	.83692	.65397	.5291	.1948	.8271	49
12	.54756	.83676	.65438	.5282	.1951	.8263	48
13	.54781	.83660	.65480	.5272	.1953	.8255	47
14	.54805	.83644	.65521	.5262	.1955	.8246	46
15	.54829	.83629	.65663	1.5252	1.1958	1.8238	45
16	.54854	.83613	.65604	.5243	.1960	.8230	44
17	.54878	.83597	.65646	.5233	.1962	.8222	43
18	.54902	.83581	.65688	.5223	.1964	.8214	42
19	.54926	.83565	.65729	.5214	.1967	.8206	41
20	.54951	.83549	.65771	1.5204	1.1969	1.8198	40
21	.54975	.83533	.65813	.5195	.1971	.8190	39
22	.54999	.83517	.65854	.5185	.1974	.8182	38
23	.55024	.83501	.65896	.5175	.1976	.8174	37
24	.55048	.83485	.65938	.5166	.1978	.8166	36
25	.55072	.83469	.65980	1.5156	1.1980	1.8158	35
26	.55097	.83453	.66021	.5147	.1983	.8150	34
27	.55121	.83437	.66063	.5137	.1985	.8142	33
28	.55145	.83421	.66105	.5127	.1987	.8134	32
29	.55169	.83405	.66147	.5118	.1990	.8126	31
30	.55194	.83388	.66188	1.5108	1.1992	1.8118	30
31	.55218	.83372	.66230	.5099	.1994	.8110	29
32	.55242	.83356	.66272	.5089	.1997	.8102	28
33	.55266	.83340	.66314	.5080	.1999	.8094	27
34	.55291	.83324	.66356	.5070	.2001	.8086	26
35	.55315	.83308	.66398	1.5061	1.2004	1.8078	25
36	.55339	.83292	.66440	.5051	.2006	.8070	24
37	.55363	.83276	.66482	.5042	.2008	.8062	23
38	.55388	.83260	.66524	.5032	.2010	.8054	22
39	.55412	.83244	.66566	.5023	.2013	.8047	21
40	.55436	.83228	.66608	1.5013	1.2015	1.8039	20
41	.55460	.83211	.66650	.5004	.2017	.8031	19
42	.55484	.83195	.66692	.4994	.2020	.8023	18
43	.55509	.83179	.66734	.4985	.2022	.8015	17
44	.55533	.83163	.66776	.4975	.2024	.8007	16
45	.55557	.83147	.66818	1.4966	1.2027	1.7999	15
46	.55581	.83131	.66860	.4957	.2029	.7992	14
47	.55605	.83115	.66902	.4947	.2031	.7984	13
48	.55629	.83098	.66944	.4938	.2034	.7976	12
49	.55654	.83082	.66986	.4928	.2036	.7968	11
50	.55678	.83066	.67028	1.4919	1.2039	1.7960	10
51	.55702	.83050	.67071	.4910	.2041	.7953	9
52	.55726	.83034	.67113	.4900	.2043	.7945	8
53	.55750	.83017	.67155	.4891	.2046	.7937	7
54	.55774	.83001	.67197	.4881	.2048	.7929	6
55	.55799	.82985	.67239	1.4872	1.2050	1.7921	5
56	.55823	.82969	.67282	.4863	.2053	.7914	4
57	.55847	.82952	.67324	.4853	.2055	.7906	3
58	.55871	.82936	.67366	.4844	.2057	.7898	2
59	.55895	.82920	.67408	.4835	.2060	.7891	1
60	.55919	.82904	.67451	1.4826	1.2062	1.7883	0

M	Cosine	Sine	Cotan.	Tan.	Cosec.	Secant	M

56°

34°

M	Sine	Cosine	Tan.	Cotan.	Secant	Cosec.	M
0	.55919	.82904	.67451	1.4826	1.2062	1.7883	60
1	.55943	.82887	.67493	.4816	.2064	.7875	59
2	.55967	.82871	.67535	.4807	.2067	.7867	58
3	.55992	.82855	.67578	.4798	.2069	.7860	57
4	.56016	.82839	.67620	.4788	.2072	.7852	56
5	.56040	.82822	.67663	1.4779	1.2074	1.7844	55
6	.56064	.82806	.67705	.4770	.2076	.7837	54
7	.56088	.82790	.67747	.4761	.2079	.7829	53
8	.56112	.82773	.67790	.4751	.2081	.7821	52
9	.56136	.82757	.67832	.4742	.2083	.7814	51
10	.56160	.82741	.67875	1.4733	1.2086	1.7806	50
11	.56184	.82724	.67917	.4724	.2088	.7798	49
12	.56208	.82708	.87960	.4714	.2091	.7791	48
13	.56232	.82692	.68002	.4705	.2093	.7783	47
14	.56256	.82675	.68045	.4696	.2095	.7776	46
15	.56280	.82659	.68087	1.4687	1.2098	1.7768	45
16	.56304	.82643	.68130	.4678	.2100	.7760	44
17	.56328	.82626	.68173	.4669	.2103	.7753	43
18	.56353	.82610	.68215	.4659	.2105	.7745	42
19	.56377	.82593	.68258	.4650	.2107	.7738	41
20	.56401	.82577	.68301	1.4641	1.2110	1.7730	40
21	.56425	.82561	.68343	.4632	.2112	.7723	39
22	.56449	.82544	.68386	.4623	.2115	.7715	38
23	.56473	.82528	.68429	.4614	.2117	.7708	37
24	.56497	.82511	.68471	.4605	.2119	.7700	36
25	.56521	.82495	.68514	1.4595	1.2122	1.7693	35
26	.56545	.82478	.68557	.4586	.2124	.7685	34
27	.56569	.82462	.68600	.4577	.2127	.7678	33
28	.56593	.82445	.68642	.4568	.2129	.7670	32
29	.56617	.82429	.68685	.4559	.2132	.7663	31
30	.56641	.82413	.68728	1.4550	1.2134	1.7655	30
31	.56664	.82396	.68771	.4541	.2136	.7648	29
32	.56688	.82380	.68814	.4532	.2139	.7640	28
33	.56712	.82363	.68857	.4523	.2141	.7633	27
34	.56736	.82347	.68899	.4514	.2144	.7625	26
35	.56760	.82330	.68942	1.4505	1.2146	1.7618	25
36	.56784	.82314	.68985	.4496	.2149	.7610	24
37	.56808	.82297	.69028	.4487	.2151	.7603	23
38	.56832	.82280	.69071	.4478	.2153	.7596	22
39	.56856	.82264	.69114	.4469	.2156	.7588	21
40	.56880	.82247	.69157	1.4460	1.2158	1.7581	20
41	.56904	.82231	.69200	.4451	.2161	.7573	19
42	.56928	.82214	.69243	.4442	.2163	.7566	18
43	.56952	.82198	.69286	.4433	.2166	.7559	17
44	.56976	.82181	.69329	.4424	.2168	.7551	16
45	.57000	.82165	.69372	1.4415	1.2171	1.7544	15
46	.57023	.82148	.69415	.4406	.2173	.7537	14
47	.57047	.82131	.69459	.4397	.2175	.7529	13
48	.57071	.82115	.69502	.4388	.2178	.7522	12
49	.57095	.82098	.69545	.4379	.2180	.7514	11
50	.57119	.82082	.69588	1.4370	1.2183	1.7507	10
51	.57143	.82065	.69631	.4361	.2185	.7500	9
52	.57167	.82048	.69674	.4352	.2188	.7493	8
53	.57191	.82032	.69718	.4343	.2190	.7485	7
54	.57214	.82015	.69761	.4335	.2193	.7478	6
55	.57238	.81998	.69804	1.4326	1.2195	1.7471	5
56	.57262	.81982	.69847	.4317	.2198	.7463	4
57	.57286	.81965	.69891	.4308	.2200	.7456	3
58	.57310	.81948	.69934	.4299	.2203	.7449	2
59	.57334	.81932	.69977	.4290	.2205	.7442	1
60	.57358	.81915	.70021	1.4281	1.2208	1.7434	0
M	Cosine	Sine	Cotan.	Tan.	Cosec.	Secant	M

55°

35°

M	Sine	Cosine	Tan.	Cotan.	Secant	Cosec.	M
0	.57358	.81915	.70021	1.4281	1.2208	1.7434	60
1	.57381	.81898	.70064	.4273	.2210	.7427	59
2	.57405	.81882	.70107	.4264	.2213	.7420	58
3	.57429	.81865	.70151	.4255	.2215	.7413	57
4	.57453	.81848	.70194	.4246	.2218	.7405	56
5	.57477	.81832	.70238	1.4237	1.2220	1.7398	55
6	.57500	.81815	.70281	.4228	.2223	.7391	54
7	.57524	.81798	.70325	.4220	.2225	.7384	53
8	.57548	.81781	.70368	.4211	.2228	.7377	52
9	.57572	.81765	.70412	.4202	.2230	.7369	51
10	.57596	.81748	.70455	1.4193	1.2233	1.7362	50
11	.57619	.81731	.70499	.4185	.2235	.7355	49
12	.57643	.81714	.70542	.4176	.2238	.7348	48
13	.57667	.81698	.70686	.4167	.2240	.7341	47
14	.57691	.81681	.70629	.4158	.2243	.7334	46
15	.57714	.81664	.70573	1.4150	1.2245	1.7327	45
16	.57738	.81647	.70717	.4141	.2248	.7319	44
17	.57762	.81630	.70760	.4132	.2250	.7312	43
18	.57786	.81614	.70804	.4123	.2253	.7305	42
19	.57809	.81597	.70848	.4115	.2255	.7298	41
20	.57833	.81580	.70891	1.4106	1.2258	1.7291	40
21	.57857	.81563	.70935	.4097	.2260	.7284	39
22	.57881	.81546	.70979	.4089	.2263	.7277	38
23	.57904	.81530	.71022	.4080	.2265	.7270	37
24	.57928	.81513	.71066	.4071	.2268	.7263	36
25	.57952	.81496	.71110	1.4063	1.2270	1.7256	35
26	.57975	.81479	.71154	.4054	.2273	.7249	34
27	.57999	.81462	.71198	.4045	.2276	.7242	33
28	.58023	.81445	.71241	.4037	.2278	.7234	32
29	.58047	.81428	.71285	.4028	.2281	.7227	31
30	.58070	.81411	.71329	1.4019	1.2283	1.7220	30
31	.58094	.81395	.71373	.4011	.2286	.7213	29
32	.58118	.81378	.71417	.4002	.2288	.7206	28
33	.58141	.81361	.71461	.3994	.2291	.7199	27
34	.58165	.81344	.71505	.3985	.2293	.7192	26
35	.58189	.81327	.71549	1.3976	1.2296	1.7185	25
36	.58212	.81310	.71593	.3968	.2298	.7178	24
37	.58236	.81293	.71637	.3959	.2301	.7171	23
38	.58259	.81276	.71681	.3951	.2304	.7164	22
39	.58283	.81259	.71725	.3942	.2306	.7157	21
40	.58307	.81242	.71769	1.3933	1.2309	1.7151	20
41	.58330	.81225	.71813	.3925	.2311	.7144	19
42	.58354	.81208	.71857	.3916	.2314	.7137	18
43	.58378	.81191	.71901	.3908	.2316	.7130	17
44	.58401	.81174	.71945	.3899	.2319	.7123	16
45	.58425	.81157	.71990	1.3891	1.2322	1.7116	15
46	.58448	.81140	.72034	.3882	.2324	.7109	14
47	.58472	.81123	.72078	.3874	.2327	.7102	13
48	.58496	.81106	.72122	.3865	.2329	.7095	12
49	.58519	.81089	.72166	.3857	.2332	.7088	11
50	.58543	.81072	.72211	1.3848	1.2335	1.7081	10
51	.58566	.81055	.72255	.3840	.2337	.7075	9
52	.58590	.81038	.72299	.3831	.2340	.7068	8
53	.58614	.81021	.72344	.3823	.2342	.7061	7
54	.58637	.81004	.72388	.3814	.2345	.7054	6
55	.58661	.80987	.72432	1.3806	1.2348	1.7047	5
56	.58684	.80970	.72477	.3797	.2350	.7040	4
57	.58708	.80953	.72521	.3789	.2353	.7033	3
58	.58731	.80936	.72565	.3781	.2355	.7027	2
59	.58755	.80919	.72610	.3772	.2358	.7020	1
60	.58778	.80902	.72654	1.3764	1.2361	1.7013	0
M	Cosine	Sine	Cotan.	Tan.	Cosec.	Secant	M

54°

36°

M	Sine	Cosine	Tan.	Cotan.	Secant	Cosec.	M
0	.58778	.80902	.72654	1.3764	1.2361	1.7013	60
1	.58802	.80885	.72699	.3755	.2363	.7006	59
2	.58825	.80867	.72743	.3747	.2366	.6999	58
3	.58849	.80850	.72788	.3738	.2368	.6993	57
4	.58873	.80833	.72832	1.3730	.2371	.6986	56
5	.58896	.80816	.72877	1.3722	1.2374	1.6979	55
6	.58920	.80799	.72921	.3713	.2376	.6972	54
7	.58943	.80782	.72966	.3705	.2379	.6965	53
8	.58967	.80765	.73010	.3697	.2382	.6959	52
9	.58990	.80747	.73055	.3688	.2384	.6952	51
10	.59014	.80730	.73100	1.3680	1.2387	1.6945	50
11	.59037	.80713	.73144	.3672	.2389	.6938	49
12	.59060	.80696	.73189	.3663	.2392	.6932	48
13	.59084	.80679	.73234	.3655	.2395	.6925	47
14	.59107	.80662	.73278	.3647	.2397	.6918	46
15	.59131	.80644	.73323	1.3638	1.2400	1.6912	45
16	.59154	.80627	.73368	.3630	.2403	.6905	44
17	.59178	.80610	.73412	.3622	.2405	.6898	43
18	.59201	.80593	.73457	.3613	.2408	.6891	42
19	.59225	.80576	.73502	.3605	.2411	.6885	41
20	.59248	.80558	.73547	1.3597	1.2413	1.6878	40
21	.59272	.80541	.73592	.3588	.2416	.6871	39
22	.59295	.80524	.73637	.3580	.2419	.6865	38
23	.59318	.80507	.73681	.3572	.2421	.6858	37
24	.59342	.80489	.73726	.3564	.2424	.6851	36
25	.59365	.80472	.73771	1.3555	1.2427	1.6845	35
26	.59389	.80455	.73816	.3547	.2429	.6838	34
27	.59412	.80437	.73861	.3539	.2432	.6831	33
28	.59435	.80420	.73906	.3531	.2435	.6825	32
29	.59459	.80403	.73951	.3522	.2437	.6818	31
30	.59482	.80386	.73996	1.3514	1.2440	1.6812	30
31	.59506	.80368	.74041	.3506	.2443	.6805	29
32	.59529	.80351	.74086	.3498	.2445	.6798	28
33	.59552	.80334	.74131	.3489	.2448	.6792	27
34	.59576	.80316	.74176	.3481	.2451	.6785	26
35	.59599	.80299	.74221	1.3473	1.2453	1.6779	25
36	.59622	.80282	.74266	.3465	.2456	.6772	24
37	.59646	.80264	.74312	.3457	.2459	.6766	23
38	.59669	.80247	.74357	.3449	.2461	.6759	22
39	.59692	.80230	.74402	.3440	.2464	.6752	21
40	.59716	.80212	.74447	1.3432	1.2467	1.6746	20
41	.59739	.80195	.74492	.3424	.2470	.6739	19
42	.59762	.80177	.74538	.3416	.2472	.6733	18
43	.59786	.80160	.74583	.3408	.2475	.6726	17
44	.59809	.80143	.74628	.3400	.2478	.6720	16
45	.59832	.80125	.74673	1.3392	1.2480	1.6713	15
46	.59856	.80108	.74719	1.3383	.2483	.6707	14
47	.59879	.80090	.74764	.3375	.2486	.6700	13
48	.59902	.80073	.74809	.3367	.2488	.6694	12
49	.59926	.80056	.74855	.3359	.2491	.6687	11
50	.59949	.80038	.74900	1.3351	1.2494	1.6681	10
51	.59972	.80021	.74946	.3343	.2497	.6674	9
52	.59995	.80003	.74991	.3335	.2499	.6668	8
53	.60019	.79986	.75037	.3327	.2502	.6661	7
54	.60042	.79968	.75082	.3319	.2505	.6655	6
55	.60065	.79951	.75128	1.3311	1.2508	1.6648	5
56	.60088	.79933	.75173	.3303	.2510	.6642	4
57	.60112	.79916	.75219	.3294	.2513	.6636	3
58	.60135	.79898	.75264	.3286	.2516	.6629	2
59	.60158	.79881	.75310	.3278	.2519	.6623	1
60	.60181	.79863	.75355	1.3270	1.2521	1.6616	0

M	Cosine	Sine	Cotan.	Tan.	Cosec.	Secant	M

53°

37°

M	Sine	Cosine	Tan.	Cotan.	Secant	Cosec.	M
0	.60181	.79863	.75355	1.3270	1.2521	1.6616	60
1	.60205	.79846	.75401	.3262	1.2524	1.6610	59
2	.60228	.79828	.75447	.3254	1.2527	1.6603	58
3	.60251	.79811	.75492	.3246	1.2530	1.6597	57
4	.60274	.79793	.75538	.3238	1.2532	1.6591	56
5	.60298	.79776	.75584	.3230	1.2535	1.6584	55
6	.60320	.79758	.75629	.3222	1.2538	1.6578	54
7	.60344	.79741	.75675	.3214	1.2541	1.6572	53
8	.60367	.79723	.75721	.3206	1.2543	1.6565	52
9	.60390	.79706	.75767	.3198	1.2546	1.6559	51
10	.60413	.79688	.75812	1.3190	1.2549	1.6552	50
11	.60437	.79670	.75858	.3182	1.2552	1.6546	49
12	.60460	.79653	.75904	.3174	1.2554	1.6540	48
13	.60483	.79635	.75950	.3166	1.2557	1.6533	47
14	.60506	.79618	.75996	.3159	1.2560	1.6527	46
15	.60529	.79600	.76042	1.3151	1.2563	1.6521	45
16	.60552	.79582	.76088	.3143	1.2565	1.6514	44
17	.60576	.79565	.76134	.3135	1.2568	1.6508	43
18	.60599	.79547	.76179	.3127	1.2571	1.6502	42
19	.60622	.79530	.76225	.3119	1.2574	1.6496	41
20	.60645	.79512	.76271	1.3111	1.2577	1.6489	40
21	.60668	.79494	.76317	.3103	1.2579	1.6483	39
22	.60691	.79477	.76364	.3095	1.2582	1.6477	38
23	.60714	.79459	.76410	.3087	1.2585	1.6470	37
24	.60737	.79441	.76456	.3079	1.2588	1.6464	36
25	.60761	.79424	.76502	1.3071	1.2591	1.6458	35
26	.60784	.79406	.76548	.3064	1.2593	1.6452	34
27	.60807	.79388	.76594	.3056	1.2596	1.6445	33
28	.60830	.79371	.76640	.3048	1.2599	1.6439	32
29	.60853	.79353	.76686	.3040	1.2602	1.6433	31
30	.60876	.79335	.76733	1.3032	1.2605	1.6427	30
31	.60899	.79318	.76779	.3024	1.2607	1.6420	29
32	.60922	.79300	.76825	.3016	1.2610	1.6414	28
33	.60945	.79282	.76871	.3009	1.2613	1.6408	27
34	.60968	.79264	.76918	.3001	1.2616	1.6402	26
35	.60991	.79247	.76964	1.2993	1.2619	1.6396	25
36	.61014	.79229	.77010	.2985	1.2622	1.6389	24
37	.61037	.79211	.77057	.2977	1.2624	1.6383	23
38	.61061	.79193	.77103	.2970	1.2627	1.6377	22
39	.61084	.79176	.77149	.2962	1.2630	1.6371	21
40	.61107	.79158	.77196	1.2954	1.2633	1.6365	20
41	.61130	.79140	.77242	.2946	1.2636	1.6359	19
42	.61153	.79122	.77289	.2938	1.2639	1.6352	18
43	.61176	.79104	.77335	.2931	1.2641	1.6346	17
44	.61199	.79087	.77382	.2923	1.2644	1.6340	16
45	.61222	.79069	.77428	1.2915	1.2647	1.6334	15
46	.61245	.79051	.77475	.2907	1.2650	1.6328	14
47	.61268	.79033	.77521	.2900	1.2653	1.6322	13
48	.61290	.79015	.77568	.2892	1.2656	1.6316	12
49	.61314	.78998	.77614	.2884	1.2659	1.6309	11
50	.61337	.78980	.77661	1.2876	1.2661	1.6303	10
51	.61360	.78962	.77708	.2869	1.2664	1.6297	9
52	.61383	.78944	.77754	.2861	1.2667	1.6291	8
53	.61405	.78926	.77801	.2853	1.2670	1.6285	7
54	.61428	.78908	.77848	.2845	1.2673	1.6279	6
55	.61451	.78890	.77895	1.2838	1.2676	1.6273	5
56	.61474	.78873	.77941	.2830	1.2679	1.6267	4
57	.61497	.78855	.77988	.2822	1.2681	1.6261	3
58	.61520	.78837	.78035	.2815	1.2684	1.6255	2
59	.61543	.78819	.78082	.2807	1.2687	1.6249	1
60	.61566	.78801	.78128	1.2799	1.2690	1.6243	0

M	Cosine	Sine	Cotan.	Tan.	Cosec.	Secant	M

52°

38°

M	Sine	Cosine	Tan.	Cotan.	Secant	Cosec.	M
0	.61566	.78801	.78128	1.2799	1.2690	1.6243	60
1	.61589	.78783	.78175	.2792	.2693	.6237	59
2	.61612	.78765	.78222	.2784	.2696	.6231	58
3	.61635	.78747	.78269	.2776	.2699	.6224	57
4	.61658	.78729	.78316	.2769	.2702	.6218	56
5	.61681	.78711	.78363	1.2761	1.2705	1.6212	55
6	.61703	.78693	.78410	.2753	.2707	.6206	54
7	.61726	.78675	.78457	.2746	.2710	.6200	53
8	.61749	.78657	.78504	.2738	.2713	.6194	52
9	.61772	.78640	.78551	.2730	.2716	.6188	51
10	.61795	.78622	.78598	1.2723	1.2719	1.6182	50
11	.61818	.78604	.78645	.2715	.2722	.6176	49
12	.61841	.78586	.78692	.2708	.2725	.6170	48
13	.61864	.78568	.78739	.2700	.2728	.6164	47
14	.61886	.78550	.78786	.2692	.2731	.6159	46
15	.61909	.78532	.78834	1.2685	1.2734	1.6153	45
16	.61932	.78514	.78881	.2677	.2737	.6147	44
17	.61955	.78496	.78928	.2670	.2739	.6141	43
18	.61978	.78478	.78975	.2662	.2742	.6135	42
19	.62001	.78460	.79022	.2655	.2745	.6129	41
20	.62023	.78441	.79070	.2647	1.2748	1.6123	40
21	.62046	.78423	.79117	.2639	.2751	.6117	39
22	.62069	.78405	.79164	.2632	.2754	.6111	38
23	.62092	.78387	.79212	.2624	.2757	.6105	37
24	.62115	.78369	.79259	.2617	.2760	.6099	36
25	.62137	.78351	.79306	1.2609	1.2763	1.6093	35
26	.62160	.78333	.79354	.2602	.2766	.6087	34
27	.62183	.78315	.79401	.2594	.2769	.6081	33
28	.62206	.78297	.79449	.2587	.2772	.6077	32
29	.62229	.78279	.79496	.2579	.2775	.6070	31
30	.62251	.78261	.79543	1.2572	1.2778	1.6064	30
31	.62274	.78243	.79591	.2564	.2781	.6058	29
32	.62297	.78224	.79639	.2557	.2784	.6052	28
33	.62320	.78206	.79686	.2549	.2787	.6046	27
34	.62342	.78188	.79734	.2542	.2790	.6040	26
35	.62365	.78170	.79781	1.2534	1.2793	1.6034	25
36	.62388	.78152	.79829	.2527	.2795	.6029	24
37	.62411	.78134	.79876	.2519	.2798	.6023	23
38	.62433	.78116	.79924	.2512	.2801	.6017	22
39	.62456	.78097	.79972	.2504	.2804	.6011	21
40	.62479	.78079	.80020	.2497	1.2807	1.6005	20
41	.62501	.78061	.80067	.2489	.2810	.6000	19
42	.62524	.78043	.80115	.2482	.2813	.5994	18
43	.62547	.78025	.80163	.2475	.2816	.5988	17
44	.62570	.78007	.80211	.2467	.2819	.5982	16
45	.62592	.77988	.80258	1.2460	1.2822	1.5976	15
46	.62615	.77970	.80306	.2452	.2825	.5971	14
47	.62638	.77952	.80354	.2445	.2828	.5965	13
48	.62660	.77934	.80402	.2437	.2831	.5959	12
49	.62683	.77915	.80450	.2430	.2834	.5953	11
50	.62706	.77897	.80498	1.2423	1.2837	1.5947	10
51	.62728	.77879	.80546	.2415	.2840	.5942	9
52	.62751	.77861	.80594	.2408	.2843	.5936	8
53	.62774	.77842	.80642	.2400	.2846	.5930	7
54	.62796	.77824	.80690	.2393	.2849	.5924	6
55	.62819	.77806	.80738	1.2386	1.2852	1.5919	5
56	.62841	.77788	.80786	.2378	.2855	.5913	4
57	.62864	.77769	.80834	.2371	.2858	.5907	3
58	.62887	.77751	.80882	.2364	.2861	.5901	2
59	.62909	.77733	.80930	.2356	.2864	.5896	1
60	.62932	.77715	.80978	1.2349	1.2867	1.5890	0
M	Cosine	Sine	Cotan.	Tan.	Cosec.	Secant	M

51°

39°

M	Sine	Cosine	Tan.	Cotan.	Secant	Cosec.	M
0	.62932	.77715	.80978	1.2349	1.2867	1.5890	60
1	.62955	.77696	.81026	.2342	.2871	.5884	59
2	.62977	.77678	.81075	.2334	.2874	.5879	58
3	.63000	.77660	.81123	.2327	.2877	.5873	57
4	.63022	.77641	.81171	.2320	.2880	.5867	56
5	.63045	.77623	.81219	1.2312	1.2883	1.5862	55
6	.63067	.77605	.81268	.2305	.2886	.5856	54
7	.63090	.77586	.81316	.2297	.2889	.5850	53
8	.63113	.77568	.81364	.2290	.2892	.5845	52
9	.63135	.77549	.81413	.2283	.2895	.5839	51
10	.63158	.77531	.81461	.2276	1.2898	1.5833	50
11	.63180	.77513	.81509	.2268	.2901	.5828	49
12	.63203	.77494	.81558	.2261	.2904	.5822	48
13	.63225	.77476	.81606	.2254	.2907	.5816	47
14	.63248	.77458	.81655	.2247	.2910	.5811	46
15	.63270	.77439	.81703	1.2239	1.2913	1.5805	45
16	.63293	.77421	.81752	.2232	.2916	.5799	44
17	.63315	.77402	.81800	.2225	.2919	.5794	43
18	.63338	.77384	.81849	.2218	.2922	.5788	42
19	.63360	.77365	.81898	.2210	.2926	.5783	41
20	.63383	.77347	.81946	1.2203	1.2929	1.5777	40
21	.63405	.77329	.81995	.2196	.2932	.5771	39
22	.63428	.77310	.82043	.2189	.2935	.5766	38
23	.63450	.77292	.82092	.2181	.2938	.5760	37
24	.63473	.77273	.82141	.2174	.2941	.5755	36
25	.63495	.77255	.82190	1.2167	1.2944	1.5749	35
26	.63518	.77236	.82238	.2160	.2947	.5743	34
27	.63540	.77218	.82287	.2152	.2950	.5738	33
28	.63563	.77199	.82336	.2145	.2953	.5732	32
29	.63585	.77181	.82385	.2138	.2956	.5727	31
30	.63608	.77162	.82434	.2131	1.2960	1.5721	30
31	.63630	.77144	.82482	.2124	.2963	.5716	29
32	.63653	.77125	.82531	.2117	.2966	.5710	28
33	.63675	.77107	.82580	.2109	.2969	.5705	27
34	.63697	.77088	.82629	.2102	.2972	.5699	26
35	.63720	.77070	.82678	1.2095	1.2975	1.5694	25
36	.63742	.77051	.82727	.2088	.2978	.5688	24
37	.63765	.77033	.82776	.2081	.2981	.5683	23
38	.63787	.77014	.82825	.2074	.2985	.5677	22
39	.63810	.76996	.82874	.2066	.2988	.5672	21
40	.63832	.76977	.82923	1.2059	1.2991	1.5666	20
41	.63854	.76958	.82972	.2052	.2994	.5661	19
42	.63877	.76940	.83022	.2045	.2997	.5655	18
43	.63899	.76921	.83071	.2038	.3000	.5650	17
44	.63921	.76903	.83120	.2031	.3003	.5644	16
45	.63944	.76884	.83169	1.2024	1.3006	1.5639	15
46	.63966	.76865	.83218	.2016	.3010	.5633	14
47	.63989	.76847	.83267	.2009	.3013	.5628	13
48	.64011	.76828	.83317	.2002	.3016	.5622	12
49	.64033	.76810	.83366	.1995	.3019	.5617	11
50	.64056	.76791	.83415	1.1988	1.3022	1.5611	10
51	.64078	.76772	.83465	.1981	.3025	.5606	9
52	.64100	.76754	.83514	.1974	.3029	.5600	8
53	.64123	.76735	.83563	.1967	.3032	.5595	7
54	.64145	.76716	.83613	.1960	.3035	.5590	6
55	.64167	.76698	.83662	1.1953	1.3038	1.5584	5
56	.64189	.76679	.83712	.1946	.3041	.5579	4
57	.64212	.76660	.83761	.1939	.3044	.5573	3
58	.64234	.76642	.83811	.1932	.3048	.5568	2
59	.64256	.76623	.83860	.1924	.3051	.5563	1
60	.64279	.76604	.83910	1.1917	1.3054	1.5557	0
M	Cosine	Sine	Cotan.	Tan.	Cosec.	Secant	M

50°

40°

M	Sine	Cosine	Tan.	Cotsn.	Secant	Cosec.	M
0	.64279	.76604	.83910	1.1917	1.3054	1.5557	60
1	.64301	.76586	.83959	.1910	.3057	.5552	59
2	.64323	.76567	.84009	.1903	.3060	.5546	58
3	.64345	.76548	.84059	.1896	.3064	.5541	57
4	.64368	.76530	.84108	.1889	.3067	.5536	56
5	.64390	.76511	.84158	1.1882	1.3070	1.5530	55
6	.64412	.76492	.84208	.1875	.3073	.5525	54
7	.64435	.76473	.84257	.1868	.3076	.5520	53
8	.64457	.76455	.84307	.1861	.3080	.5514	52
9	.64479	.76436	.84357	.1854	.3083	.5509	51
10	.64501	.76417	.84407	1.1847	1.3086	1.5503	50
11	.64523	.76398	.84457	.1840	.3089	.5498	49
12	.64546	.76380	.84506	.1833	.3092	.5493	48
13	.64568	.76361	.84556	.1826	.3096	.5487	47
14	.64590	.76342	.84606	.1819	.3099	.5482	46
15	.64612	.76323	.84656	1.1812	1.3102	1.5477	45
16	.64635	.76304	.84706	.1805	.3105	.5471	44
17	.64657	.76286	.84756	.1798	.3109	.5466	43
18	.64679	.76267	.84806	.1791	.3112	.5461	42
19	.64701	.76248	.84856	.1785	.3115	.5456	41
20	.64723	.76229	.84906	1.1778	1.3118	1.5450	40
21	.64745	.76210	.84956	.1771	.3121	.5445	39
22	.64768	.76191	.85006	.1764	.3125	.5440	38
23	.64790	.76173	.85056	.1757	.3128	.5434	37
24	.64812	.76154	.85107	.1750	.3131	.5429	36
25	.64834	.76135	.85157	1.1743	1.3134	1.5424	35
26	.64856	.76116	.85207	.1736	.3138	.5419	34
27	.64878	.76097	.85257	.1729	.3141	.5413	33
28	.64900	.76078	.85307	.1722	.3144	.5408	32
29	.64923	.76059	.85358	.1715	.3148	.5403	31
30	.64945	.76041	.85408	1.1708	1.3151	1.5398	30
31	.64967	.76022	.85458	.1702	.3154	.5392	29
32	.64989	.76003	.85509	.1695	.3157	.5387	28
33	.65011	.75984	.85559	.1688	.3161	.5382	27
34	.65033	.75965	.85609	.1681	.3164	.5377	26
35	.65055	.75946	.85660	1.1674	1.3167	1.5371	25
36	.65077	.75927	.85710	.1667	.3170	.5366	24
37	.65100	.75908	.85761	.1660	.3174	.5361	23
38	.65121	.75889	.85811	.1653	.3177	.5356	22
39	.65144	.75870	.85862	.1647	.3180	.5351	21
40	.65166	.75851	.85912	1.1640	1.3184	1.5345	20
41	.65188	.75832	.85963	.1633	.3187	.5340	19
42	.65210	.75813	.86013	.1626	.3190	.5335	18
43	.65232	.75794	.86064	.1619	.3193	.5330	17
44	.65254	.75775	.86115	.1612	.3197	.5325	16
45	.65276	.75756	.86165	1.1605	1.3200	1.5319	15
46	.65298	.75737	.86216	.1599	.3203	.5314	14
47	.65320	.75718	.86267	.1592	.3207	.5309	13
48	.65342	.75700	.86318	.1585	.3210	.5304	12
49	.65364	.75680	.86368	.1578	.3213	.5290	11
50	.65386	.75661	.86419	1.1571	1.3217	1.5294	10
51	.65408	.75642	.86470	.1565	.3220	.5289	9
52	.65430	.75623	.86521	.1558	.3223	.5283	8
53	.65452	.75604	.86572	.1551	.3227	.5278	7
54	.65474	.75585	.86623	.1544	.3230	.5273	6
55	.65496	.75566	.86674	1.1537	1.3233	1.5268	5
56	.65518	.75547	.86725	.1531	.3237	.5263	4
57	.65540	.75528	.86775	.1524	.3240	.5258	3
58	.65562	.75509	.86828	.1517	.3243	.5253	2
59	.65584	.75490	.86878	.1510	.3247	.5248	1
60	.65606	.75471	.86929	1.1504	1.3250	1.5242	0
M	Cosine	Sine	Cotan.	Tan.	Cosec.	Secant	M

49°

41°

M	Sine	Cosine	Tan.	Cotan.	Secant	Cosec.	M
0	.65606	.75471	.86929	1.1504	1.3250	1.5242	60
1	.65628	.75452	.86980	.1497	.3253	.5237	59
2	.65650	.75433	.87031	.1490	.3257	.5232	58
3	.65672	.75414	.87082	.1483	.3260	.5227	57
4	.65694	.75394	.87133	.1477	.3263	.5222	56
5	.65716	.75375	.87184	.1470	1.3267	1.5217	55
6	.65737	.75356	.87235	.1463	.3270	.5212	54
7	.65759	.75337	.87287	.1456	.3274	.5207	53
8	.65781	.75318	.87338	.1450	.3277	.5202	52
9	.65803	.75299	.87389	.1443	.3280	.5197	51
10	.65825	.75280	.87441	1.1436	1.3284	1.5192	50
11	.65847	.75261	.87492	.1430	.3287	.5187	49
12	.65869	.75241	.87543	.1423	.3290	.5182	48
13	.65891	.75222	.87595	.1416	.3294	.5177	47
14	.65913	.75203	.87646	.1409	.3297	.5171	46
15	.65934	.75184	.87698	1.1403	1.3301	1.5166	45
16	.65956	.75165	.87749	.1396	.3304	.5161	44
17	.65978	.75146	.87801	.1389	.3307	.5156	43
18	.66000	.75126	.87852	.1383	.3311	.5151	42
19	.66022	.75107	.87904	.1376	.3314	.5146	41
20	.66044	.75088	.87955	1.1369	1.3318	1.5141	40
21	.66066	.75069	.88007	.1363	.3321	.5136	39
22	.66087	.75049	.88058	.1356	.3324	.5131	38
23	.66109	.75030	.88110	.1349	.3328	.5126	37
24	.66131	.75011	.88162	.1343	.3331	.5121	36
25	.66153	.74992	.88213	1.1336	1.3335	1.5116	35
26	.66175	.74973	.88265	.1329	.3338	.5111	34
27	.66197	.74953	.88317	.1323	.3342	.5106	33
28	.66218	.74934	.88369	.1316	.3345	.5101	32
29	.66240	.74915	.88421	.1309	.3348	.5096	31
30	.66262	.74895	.88472	1.1303	1.3352	1.5092	30
31	.66284	.74876	.88524	.1296	.3355	.5087	29
32	.66305	.74857	.88576	.1290	.3359	.5082	28
33	.66327	.74838	.88628	.1283	.3362	.5077	27
34	.66349	.74818	.88680	.1276	.3366	.5072	26
35	.66371	.74799	.88732	1.1270	1.3369	1.5067	25
36	.66393	.74780	.88784	.1263	.3372	.5062	24
37	.66414	.74760	.88836	.1257	.3376	.5057	23
38	.66436	.74741	.88888	.1250	.3379	.5052	22
39	.66458	.74722	.88940	.1243	.3383	.5047	21
40	.66479	.74702	.88992	1.1237	1.3386	1.5042	20
41	.66501	.74683	.89044	.1230	.3390	.5037	19
42	.66523	.74664	.89097	.1224	.3393	.5032	18
43	.66545	.74644	.89149	.1217	.3397	.5027	17
44	.66566	.74625	.89201	.1211	.3400	.5022	16
45	.66588	.74606	.89253	1.1204	1.3404	1.5018	15
46	.66610	.74586	.89306	.1197	.3407	.5013	14
47	.66631	.74567	.89358	.1191	.3411	.5008	13
48	.66653	.74548	.89410	.1184	.3414	.5003	12
49	.66675	.74528	.89463	.1178	.3418	.4998	11
50	.66697	.74509	.89515	1.1171	1.3421	1.4993	10
51	.66718	.74489	.89567	.1165	.3425	.4988	9
52	.66740	.74470	.89620	.1158	.3428	.4983	8
53	.66762	.74450	.89672	.1152	.3432	.4979	7
54	.66783	.74431	.89725	.1145	.3435	.4974	6
55	.66805	.74412	.89777	1.1139	1.3439	1.4969	5
56	.66826	.74392	.89830	.1132	.3442	.4964	4
57	.66848	.74373	.89882	.1126	.3446	.4959	3
58	.66870	.74353	.89935	.1119	.3449	.4954	2
59	.66891	.74334	.89988	.1113	.3453	.4949	1
60	.66913	.74314	.90040	1.1106	1.3456	1.4945	0
M	Cosine	Sine	Cotan.	Tan.	Cosec.	Secant	M

48°

M	Sine	Cosine	Tan.	Cotan.	Secant	Cosec.	M
0	.66913	.74314	.90040	1.1106	1.3456	1.4945	60
1	.66935	.74295	.90093	.1100	.3460	.4940	59
2	.66956	.74275	.90146	.1093	.3463	.4935	58
3	.66978	.74256	.90198	.1086	.3467	.4930	57
4	.66999	.74236	.90251	.1080	.3470	.4925	56
5	.67021	.74217	.90304	.1074	1.3474	1.4921	55
6	.67043	.74197	.90357	.1067	.3477	.4916	54
7	.67064	.74178	.90410	.1061	.3481	.4911	53
8	.67086	.74158	.90463	.1054	.3485	.4906	52
9	.67107	.74139	.90515	.1048	.3488	.4901	51
10	.67129	.74119	.90568	.1041	1.3492	1.4897	50
11	.67150	.74100	.90621	.1035	.3495	.4892	49
12	.67172	.74080	.90674	.1028	.3499	.4887	48
13	.67194	.74061	.90727	.1022	.3502	.4882	47
14	.67215	.74041	.90780	.1015	.3506	.4877	46
15	.67237	.74022	.90834	.1009	1.3509	1.4873	45
16	.67258	.74002	.90887	.1003	.3513	.4868	44
17	.67280	.73983	.90940	.0996	.3517	.4863	43
18	.67301	.73963	.90993	.0990	.3520	.4858	42
19	.67323	.73943	.91046	.0983	.3524	.4854	41
20	.67344	.73924	.91099	1.0977	1.3527	1.4849	40
21	.67366	.73904	.91153	.0971	.3531	.4844	39
22	.67387	.73885	.91206	.0964	.3534	.4839	38
23	.67409	.73865	.91259	.0958	.3538	.4835	37
24	.67430	.73845	.91312	.0951	.3542	.4830	36
25	.67452	.73826	.91366	1.0945	1.3545	1.4825	35
26	.67473	.73806	.91419	.0939	.3549	.4821	34
27	.67495	.73787	.91473	.0932	.3552	.4816	33
28	.67516	.73767	.91526	.0926	.3556	.4811	32
29	.67537	.73747	.91580	.0919	.3560	.4806	31
30	.67559	.73728	.91633	1.0913	1.3563	1.4802	30
31	.67580	.73708	.91687	.0907	.3567	.4797	29
32	.67602	.73688	.91740	.0900	.3571	.4792	28
33	.67623	.73669	.91794	.0894	.3574	.4788	27
34	.67645	.73649	.91847	.0888	.3578	.4783	26
35	.67666	.73629	.91901	1.0881	1.3581	1.4778	25
36	.67688	.73610	.91955	.0875	.3585	.4774	24
37	.67709	.73590	.92008	.0868	.3589	.4769	23
38	.67730	.73570	.92062	.0862	.3592	.4764	22
39	.67752	.73551	.92116	.0856	.3596	.4760	21
40	.67773	.73531	.92170	1.0849	1.3600	1.4755	20
41	.67794	.73511	.92223	.0843	.3603	.4750	19
42	.67816	.73491	.92277	.0837	.3607	.4746	18
43	.67837	.73472	.92331	.0830	.3611	.4741	17
44	.67859	.73452	.92385	.0824	.3614	.4736	16
45	.67880	.73432	.92439	1.0818	1.3618	1.4732	15
46	.67901	.73412	.92493	.0812	.3622	.4727	14
47	.67923	.73393	.92547	.0805	.3625	.4723	13
48	.67944	.73373	.92601	.0799	.3629	.4718	12
49	.67965	.73353	.92655	.0793	.3633	.4713	11
50	.67987	.73333	.92709	1.0786	1.3636	1.4709	10
51	.68008	.73314	.92763	.0780	.3640	.4704	9
52	.68029	.73294	.92817	.0774	.3644	.4699	8
53	.68051	.73274	.92871	.0767	.3647	.4695	7
54	.68072	.73254	.92926	.0761	.3651	.4690	6
55	.68093	.73234	.92980	1.0755	1.3655	1.4686	5
56	.68115	.73215	.93034	.0749	.3658	.4681	4
57	.68136	.73195	.93088	.0742	.3662	.4676	3
58	.68157	.73175	.93143	.0736	.3666	.4672	2
59	.68178	.73155	.93197	.0730	.3669	.4667	1
60	.68200	.73135	.93251	1.0724	1.3673	1.4663	0

M	Cosine	Sine	Cotan.	Tan.	Cosec.	Secant	M

47°

71

43°

M	Sine	Cosine	Tan.	Cotan.	Secant	Cosec.	M
0	.68200	.73135	.93251	1.0724	1.3673	1.4663	60
1	.68221	.73115	.93306	.0717	.3677	.4658	59
2	.68242	.73096	.93360	.0711	.3681	.4654	58
3	.68264	.73076	.93415	.0705	.3684	.4649	57
4	.68285	.73056	.93469	.0699	.3688	.4644	56
5	.68306	.73036	.93524	1.0692	1.3692	1.4640	55
6	.68327	.73016	.93578	.0686	.3695	.4635	54
7	.68349	.72996	.93633	.0680	.3699	.4631	53
8	.68370	.72976	.93687	.0674	.3703	.4626	52
9	.68391	.72956	.93742	.0667	.3707	.4622	51
10	.68412	.72937	.93797	.0661	1.3710	1.4617	50
11	.68433	.72917	.93851	.0655	.3714	.4613	49
12	.68455	.72897	.93906	.0649	.3718	.4608	48
13	.68476	.72877	.93961	.0643	.3722	.4604	47
14	.68497	.72857	.94016	.0636	.3725	.4599	46
15	.68518	.72837	.94071	1.0630	1.3729	1.4595	45
16	68539	.72817	.94125	.0624	.3733	.4590	44
17	.68561	.72797	.94180	.0618	.3737	.4586	43
18	.68582	.72777	.94235	.0612	.3740	.4581	42
19	.68603	.72757	.94290	.0605	.3744	.4577	41
20	.68624	.72737	.94345	.0599	1.3748	1.4572	40
21	.68645	.72717	.94400	.0593	.3752	.4568	39
22	.68666	.72697	.94455	.0587	.3756	.4563	38
23	.68688	.72677	.94510	.0581	.3759	.4559	37
24	.68709	.72657	.94565	.0575	.3763	.4554	36
25	.68730	.72637	.94620	1.0568	1.3767	1.4550	35
26	.68751	.72617	.94675	.0562	.3771	.4545	34
27	.68772	.72597	.94731	.0556	.3774	.4541	33
28	.68793	.72577	.94786	.0550	.3778	.4536	32
29	.68814	.72557	.94841	.0544	.3782	.4532	31
30	.68835	.72537	.94896	.0538	1.3786	1.4527	30
31	.68856	.72517	.94952	.0532	.3790	.4523	29
32	.68878	.72497	.95007	.0525	.3794	.4518	28
33	.68899	.72477	.95062	.0519	.3797	.4514	27
34	.68920	.72457	.95118	.0513	.3801	.4510	26
35	.68941	.72437	.95173	1.0507	1.3805	1.4505	25
36	.68962	.72417	.95229	.0501	.3809	.4501	24
37	.68983	.72397	.95284	.0495	.3813	.4496	23
38	.69004	.72377	.95340	.0489	.3816	.4492	22
39	.69025	.72357	.95395	.0483	.3820	.4487	21
40	.69046	.72337	.95451	1.0476	1.3824	1.4483	20
41	.69067	.72317	.95506	.0470	.3828	.4479	19
42	.69088	.72297	.95562	.0464	.3832	.4474	18
43	.69109	.72277	.95618	.0458	.3836	.4470	17
44	.69130	.72256	.95673	.0452	.3839	.4465	16
45	.69151	.72236	.95729	1.0446	1.3843	1.4461	15
46	.69172	.72216	.95785	.0440	.3847	.4457	14
47	.69193	.72196	.95841	.0434	.3851	.4452	13
48	.69214	.72176	.95896	.0428	.3855	.4448	12
49	.69235	.72156	.95952	.0422	.3859	.4443	11
50	.69256	.72136	.96008	1.0416	1.3863	1.4439	10
51	.69277	.72115	.96064	.0410	.3867	.4435	9
52	.69298	.72095	.96120	.0404	.3870	.4430	8
53	.69319	.72075	.96176	.0397	.3874	.4426	7
54	.69340	.72055	.96232	.0391	.3878	.4422	6
55	.69361	.72035	.96288	1.0385	1.3882	1.4417	5
56	.69382	.72015	.96344	.0379	.3886	.4413	4
57	.69403	.71994	.96400	.0373	.3890	.4408	3
58	.69424	.71974	.96456	.0367	.3894	.4404	2
59	.69445	.71954	.96513	.0361	.3898	.4400	1
60	.69466	.71934	.96569	1.0355	1.3902	1.4395	0
M	Cosine	Sine	Cotan.	Tan.	Cosec.	Secant	M

46°

44°

M	Sine	Cosine	Tan.	Cotan.	Secant	Cosec.	M
0	.69466	.71934	.96569	1.0355	1.3902	1.4395	60
1	.69487	.71914	.96625	.0349	.3905	.4391	59
2	.69508	.71893	.96681	.0343	.3909	.4387	58
3	.69528	.71873	.96738	.0337	.3913	.4382	57
4	.69549	.71853	.96794	.0331	.3917	.4378	56
5	.69570	.71833	.96850	1.0325	1.3921	1.4374	55
6	.69591	.71813	.96907	.0319	.3925	.4370	54
7	.69612	.71792	.96963	.0313	.3929	.4365	53
8	.69633	.71772	.97020	.0307	.3933	.4361	52
9	.69654	.71752	.97076	.0301	.3937	.4357	51
10	.69675	.71732	.97133	1.0295	1.3941	1.4352	50
11	.69696	.71711	.97189	.0289	.3945	.4348	49
12	.69716	.71691	.97246	.0283	.3949	.4344	48
13	.69737	.71671	.97302	.0277	.3953	.4339	47
14	.69758	.71650	.97359	.0271	.3957	.4335	46
15	.69779	.71630	.97416	1.0265	1.3960	1.4331	45
16	.69800	.71610	.97472	.0259	.3964	.4327	44
17	.69821	.71589	.97529	.0253	.3968	.4322	43
18	.69841	.71569	.97586	.0247	.3972	.4318	42
19	.69862	.71549	.97643	.0241	.3976	.4314	41
20	.69883	.71529	.97700	1.0235	1.3980	1.4310	40
21	.69904	.71508	.97756	.0229	.3984	.4305	39
22	.69925	.71488	.97813	.0223	.3988	.4301	38
23	.69945	.71468	.97870	.0218	.3992	.4297	37
24	.69966	.71447	.97927	.0212	.3996	.4292	36
25	.69987	.71427	.97984	1.0206	1.4000	1.4288	35
26	.70008	.71406	.98041	.0200	.4004	.4284	34
27	.70029	.71386	.98098	.0194	.4008	.4280	33
28	.70049	.71366	.98155	.0188	.4012	.4276	32
29	.70070	.71345	.98212	.0182	.4016	.4271	31
30	.70091	.71325	.98270	1.0176	1.4020	1.4267	30
31	.70112	.71305	.98327	.0170	.4024	.4263	29
32	.70132	.71284	.98384	.0164	.4028	.4259	28
33	.70153	.71264	.98441	.0158	.4032	.4254	27
34	.70174	.71243	.98499	.0152	.4036	.4250	26
35	.70194	.71223	.98556	1.0146	1.4040	1.4246	25
36	.70215	.71203	.98613	.0141	.4044	.4242	24
37	.70236	.71182	.98671	.0135	.4048	.4238	23
38	.70257	.71162	.98728	.0129	.4052	.4233	22
39	.70277	.71141	.98786	.0123	.4056	.4229	21
40	.70298	.71121	.98843	1.0117	1.4060	1.4225	20
41	.70319	.71100	.98901	.0111	.4065	.4221	19
42	.70339	.71080	.98958	.0105	.4069	.4217	18
43	.70360	.71059	.99016	.0099	.4073	.4212	17
44	.70381	.71039	.99073	.0093	.4077	.4208	16
45	.70401	.71018	.99131	1.0088	1.4081	1.4204	15
46	.70422	.70998	.99189	.0082	.4085	.4200	14
47	.70443	.70977	.99246	.0076	.4089	.4196	13
48	.70463	.70957	.99304	.0070	.4093	.4192	12
49	.70484	.70936	.99362	.0064	.4097	.4188	11
50	.70505	.70916	.99420	1.0058	1.4101	1.4183	10
51	.70525	.70895	.99478	.0052	.4105	.4179	9
52	.70546	.70875	.99536	.0047	.4109	.4175	8
53	.70566	.70854	.99593	.0041	.4113	.4171	7
54	.70587	.70834	.99651	.0035	.4117	.4167	6
55	.70608	.70813	.99709	1.0029	1.4122	1.4163	5
56	.70628	.70793	.99767	.0023	.4126	.4159	4
57	.70649	.70772	.99826	.0017	.4130	.4154	3
58	.70669	.70752	.99884	.0012	.4134	.4150	2
59	.70690	.70731	.99942	.0006	.4138	.4146	1
60	.70711	.70711	1.0000	1.0000	1.4142	1.4142	0
M	Cosine	Sine	Cotan.	Tan.	Cosec.	Secant	M

45°

Grinding of Twist Drills

Dipping High Speed Drills in Water is Bad Practice

In up-to-date drilling practice drilling different grades of materials requires at times a modification of the commercial 118° drill point for maximum results.

Hard materials require a blunter point with the more acute angle for softer materials.

ANGLE OF POINTS

		Point
Fig. 1, 2	Average Class of Work	118° included angle 12° to 15° lip clearance
Fig. 3	Alloy Steels, Monel Metal, Stainless Steel, Heat Treated Steels, Drop Forgings (Automobile Connecting Rods) Brinell Hardness No. 240	125° included angle 10° to 12° lip clearance
Fig. 4	Soft and Medium Cast Iron, Aluminum, Marble, Slate, Plastics, Wood, Hard Rubber, Bakelite, Fibre	90° to 130° included angle 12° lip clearance Flat cutting lip for marble
Fig. 5	Copper, Soft and Medium Hard Brass Magnesium Alloys	100° to 118° included angle 12° to 15° lip clearance 60° to 118° included angle 15° lip clearance Slightly flat face of cutting lips reducing rake angle to 5°
Fig. 6	Wood, Rubber, Bakelite, Fibre, Aluminum, Die Castings, Plastics	60° included angle 12° to 15° lip clearance
Fig. 7	Steel 7% to 13% Manganese, Tough Alloy Steels, Armor Plate and hard materials	150° included angle 7° to 10° lip clearance Slightly flat face of cutting lips
Fig. 8	Brass, Soft Bronze	118° included angle 12° to 15° lip clearance Slightly flat face of cutting lips
Fig. 9	Crankshafts, Deep Holes in Soft Steel, Hard Steel, Cast Iron, Nickel and Manganese Alloys	118° included angle Chisel Point 9° lip clearance
Fig. 10	Thin Sheet Metal, Copper, Fibre, Plastics, Wood	−5° to +12° lip angles For drills over ¼″ diameter make angle of bit point to suit work

Grinding of Twist Drills

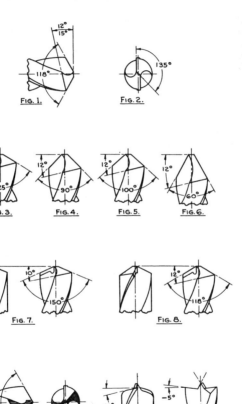

FIG. 1.

FIG. 2.

FIG. 3.

FIG. 4.

FIG. 5.

FIG. 6.

FIG. 7.

FIG. 8.

FIG. 9.

FIG. 10.

DRILLING SPEEDS FOR HIGH SPEED STEEL DRILLS
expressed in
Feet per minute, meters per second, and meters per minute.

Material	Suggested Cutting Speeds		
	ft./min.	m/sec.	m/min.
Aluminum and its Alloys................	200-300	1.0 -1.5	60- 90
Brass and Bronze (ordinary)...........	150-300	.75 -1.5	45- 90
Bronze (High Tensile)................	70-150	.35 - .75	20- 45
Die Castings (Zinc Base)..............	300-400	1.5 -2.0	90-120
Iron—Cast (soft)......................	100-150	.5 - .75	30- 45
Cast (medium hard)..............	70-100	.35 - .5	20- 30
Hard Chilled....................	30- 40	.15 - .2	10- 12
Malleable......................	80- 90	.4 - .45	25- 30
Magnesium and its Alloys.............	250-400	1.25 -2.0	75-120
Monel Metal or High-Nickel Steel......	30- 50	.15 - .25	10- 15
Plastics or Similar Materials (Bakelite).	100-300	.5 -1.5	30- 90
Steel—Mild .2 carbon to .3 carbon	80-110	.4 - .55	25- 35
Steel .4 carbon to .5 carbon	70- 80	.35 - .4	20- 25
Tool 1.2 carbon	50- 60	.25 - .3	15- 20
Forgings......................	40- 50	.2 - .25	12- 15
Alloy—300 to 400 Brinell.....	20- 30	.1 - .15	5- 10
High Tensile (Heat Treated)			
35 to 40 Rockwell C.............	30- 40	.15 - .2	10- 12
40 to 45 Rockwell C.............	25- 35	.125- .175	8- 10
45 to 50 Rockwell C.............	15- 25	.075- .125	5- 8
50 to 55 Rockwell C.............	7- 15	.035- .075	2- 5
Stainless Steel			
Free Machining Grades..............	30- 80	.15 - .4	10- 25
Work Hardening Grades.............	15- 50	.075- .25	5- 15
Titanium Alloy Sheet....................	50- 60	.25 - .3	15- 20
Titanium Alloys			
Ti—75A (Commercially Pure).........	50- 60	.25 - .3	15- 20
Ti-6AL-4VA................ BR. 310-350	20- 35	.1 - .175	5- 10
Inconel Alloy........... BR. 200-400	15- 20	.075- .1	4- 6
Hastelloy (Wrought).... BR. 140-310	15- 20	.075- .1	4- 6
Hastelloy (Cast)........ BR. 200-375	5- 7	.025- .035	1- 2
Rene.................... BR. 225-400	12- 20	.06 - .1	4- 6
Zirconium Alloys........ BR. 140-280	55	- .275	15
Wood..................................	300-400	1.5 -2.0	90-120

FEEDS FOR DRILLS

Dia. of drill (inches)	Feed (inches/rev.)	Dia. of drill (mm)	Feed (mm/rev.)
Under 1/8	.001 to .002	Under 3.0	.025 to .050
1/8 to 1/4	.002 to .004	3 to 6.5	.050 to .100
1/4 to 1/2	.004 to .007	6.5 to 13.0	.100 to .175
1/2 to 1	.007 to .015	13 to 25	.175 to .380
1 and over	.015 to .025	25 and over	.380 to .650

METRIC SIZE DRILLS

With

Decimal Inch Equivalent

Diameter		Diameter		Diameter		Diameter	
mm	Inch	mm	Inch	mm	Inch	mm	Inch
0.35	.0138	2.15	.0846	4.90	.1929	7.90	.3110
0.40	.0158	2.20	.0866	5.00	.1968	8.00	.3150
0.45	.0177	2.25	.0886	5.10	.2008	8.10	.3189
0.50	.0197	2.30	.0906	5.20	.2047	8.20	.3228
0.55	.0217	2.35	.0925	5.25	.2067	8.25	.3248
0.60	.0236	2.40	.0945	5.30	.2087	8.30	.3268
0.65	.0256	2.45	.0965	5.40	.2126	8.40	.3307
0.70	.0276	2.50	.0984	5.50	.2165	8.50	.3346
0.75	.0295	2.60	.1024	5.60	.2205	8.60	.3386
0.80	.0315	2.70	.1063	5.70	.2244	8.70	.3425
0.85	.0335	2.75	.1083	5.75	.2264	8.75	.3445
0.90	.0354	2.80	.1102	5.80	.2283	8.80	.3465
0.95	.0374	2.90	.1142	5.90	.2323	8.90	.3504
1.00	.0394	3.00	.1181	6.00	.2362	9.00	.3543
1.05	.0413	3.10	.1220	6.10	.2402	9.20	.3583
1.10	.0433	3.20	.1260	6.20	.2441	9.20	.3622
1.15	.0453	3.25	.1280	6.25	.2461	9.25	.3642
1.20	.0472	3.30	.1299	6.30	.2480	9.30	.3661
1.25	.0492	3.40	.1339	6.40	.2520	9.40	.3701
1.30	.0512	3.50	.1378	6.50	.2559	9.50	.3740
1.35	.0531	3.60	.1417	6.60	.2598	9.60	.3780
1.40	.0551	3.70	.1457	6.70	.2638	9.70	.3819
1.45	.0571	3.75	.1476	6.75	.2657	9.75	.3839
1.50	.0591	3.80	.1496	6.80	.2677	9.80	.3858
1.55	.0610	3.90	.1535	6.90	.2717	9.90	.3898
1.60	.0630	4.00	.1575	7.00	.2756	10.00	.3937
1.65	.0650	4,10	.1614	7.10	.2795	10.50	.4134
1.70	.0669	4.20	.1654	7.20	.2835	11.00	.4331
1.75	.0689	4.25	.1673	7.25	.2854	11.50	.4528
1.80	.0709	4.30	.1693	7.30	.2874	12.00	.4724
1.85	.0728	4.40	.1739	7.40	.2913	12.50	.4921
1.90	.0748	4.50	.1772	7.50	.2953	13.00	.5118
1.95	.0768	4.60	.1811	7.60	.2992	13.50	.5314
2.00	.0787	4.70	.1850	7.70	.3031	14.00	.5511
2.05	.0807	4.75	.1870	7.75	.3051	14.50	.5708
2.10	.0827	4.80	.1890	7.80	.3071	15.00	.5905

TABLE OF CUTTING SPEEDS
Metric size drills

m/min. Dia. in mm	10	20	30	40	50	60	70	80	90	100	120
	Revolutions per minute										
1.5	2123	4246	6369	8493	10616	12739	14862	16985	19108	21231	25478
2	1592	3185	4777	6369	7962	9554	11146	12739	14331	15924	19108
2.5	1274	2548	3822	5096	6369	7643	8917	10191	11465	12739	15287
3	1062	2123	3185	4246	5308	6369	7431	8493	9554	10616	12739
3.5	910	1820	2730	3640	4550	5460	6369	7279	8189	9099	10919
4	796	1592	2389	3185	3981	4777	5573	6369	7166	7962	9554
4.5	708	1415	2123	2831	3539	4246	4954	5662	6369	7077	8493
6	531	1062	1592	2123	2654	3185	3715	4246	4777	5308	6369
7.5	425	849	1274	1698	2123	2548	2972	3397	3822	4246	5096
9	354	708	1062	1415	1769	2123	2477	2831	3185	3539	4246
10.5	303	607	910	1213	1517	1820	2123	2426	2730	3033	3640
12	265	531	796	1062	1327	1592	1858	2123	2389	2654	3185
15	212	425	637	849	1062	1274	1486	1698	1911	2123	2548
20	159	318	478	637	796	955	1115	1274	1433	1592	1911
25	127	255	382	510	637	764	892	1019	1146	1274	1529
30	106	212	318	425	531	637	743	849	955	1062	1274
35	91	182	273	364	455	546	637	728	819	910	1092
40	80	159	239	318	398	478	557	637	716	796	955
45	71	142	212	283	354	425	495	566	637	708	849
50	64	127	191	255	318	382	446	510	573	637	764
55	58	116	174	232	290	347	405	463	521	579	695
60	53	106	159	212	265	318	371	425	478	531	657
65	49	98	147	196	245	294	343	392	441	490	588
70	45	91	136	182	227	273	318	364	409	455	546

Table of Cutting Speeds

(Fraction Size Drills)

Feet per Min.	30'	40'	50'	60'	70'	80'	90'	100'	110'	120'	130'	140'	150'
Diameter Inches	Revolutions per Minute												
1/16	1833	2445	3056	3667	4278	4889	5500	6111	6722	7334	7945	8556	9167
1/8	917	1222	1528	1833	2139	2445	2750	3056	3361	3667	3973	4278	4584
3/16	611	815	1019	1222	1426	1630	1833	2037	2241	2445	2648	2852	3056
1/4	458	611	764	917	1070	1222	1375	1528	1681	1833	1986	2139	2292
5/16	367	489	611	733	856	978	1100	1222	1345	1467	1589	1711	1833
3/8	306	407	509	611	713	815	917	1019	1120	1222	1324	1426	1528
7/16	262	349	437	524	611	698	786	873	960	1048	1135	1222	1310
1/2	229	306	382	458	535	611	688	764	840	917	993	1070	1146
5/8	183	244	306	367	428	489	550	611	672	733	794	856	917
3/4	153	203	255	306	357	407	458	509	560	611	662	713	764
7/8	131	175	218	262	306	349	393	436	480	524	568	611	655
1	115	153	191	229	267	306	344	382	420	458	497	535	573
1 1/8	102	136	170	204	238	272	306	340	373	407	441	475	509
1 1/4	92	122	153	183	214	244	275	306	336	367	397	428	458
1 3/8	83	111	139	167	194	222	250	278	306	333	361	389	417
1 1/2	76	102	127	153	178	204	229	255	280	306	331	357	382
1 5/8	70	94	117	141	165	188	212	235	259	282	306	329	353
1 3/4	65	87	109	131	153	175	196	218	240	262	284	306	327
1 7/8	61	81	102	122	143	163	183	204	224	244	265	285	306
2	57	76	95	115	134	153	172	191	210	229	248	267	287
2 1/4	51	68	85	102	119	136	153	170	187	204	221	238	255
2 1/2	46	61	76	92	107	122	137	153	168	183	199	214	229
2 3/4	42	56	69	83	97	111	125	139	153	167	181	194	208
3	38	51	64	76	89	102	115	127	140	153	166	178	191

Table of Cutting Speeds
(Number Size Drills)

Feet Per Min.	30'	40'	50'	60'	70'	80'	90'	100'
No. Size	Revolutions per Minute							
1	503	670	838	1005	1173	1340	1508	1675
2	518	691	864	1037	1210	1382	1555	1728
3	538	717	897	1076	1255	1434	1614	1793
4	548	731	914	1097	1280	1462	1645	1828
5	558	744	930	1115	1301	1487	1673	1859
6	562	749	936	1123	1310	1498	1685	1872
7	570	760	950	1140	1330	1520	1710	1900
8	576	768	960	1151	1343	1535	1727	1919
9	585	780	975	1169	1364	1559	1754	1949
10	592	790	987	1184	1382	1579	1777	1974
11	600	800	1000	1200	1400	1600	1800	2000
12	606	808	1010	1213	1415	1617	1819	2021
13	620	826	1032	1239	1450	1652	1859	2065
14	630	840	1050	1259	1469	1679	1889	2099
15	638	851	1064	1276	1489	1702	1914	2127
16	647	863	1079	1295	1511	1726	1942	2158
17	662	883	1104	1325	1546	1766	1987	2208
18	678	904	1130	1356	1582	1808	2034	2260
19	690	920	1151	1381	1611	1841	2071	2301
20	712	949	1186	1423	1660	1898	2135	2372
21	721	961	1201	1441	1681	1922	2162	2402
22	730	973	1217	1460	1703	1946	2190	2433
23	744	992	1240	1488	1736	1984	2232	2480
24	754	1005	1257	1508	1759	2010	2262	2513
25	767	1022	1276	1533	1789	2044	2300	2555
26	779	1039	1299	1559	1819	2078	2338	2598
27	796	1061	1327	1592	1857	2122	2388	2653
28	816	1088	1360	1631	1903	2175	2447	2719
29	843	1124	1405	1685	1966	2247	2528	2809
30	892	1189	1487	1784	2081	2378	2676	2973
31	955	1273	1592	1910	2228	2546	2865	3183
32	988	1317	1647	1976	2305	2634	2964	3293
33	1014	1352	1690	2028	2366	2704	3042	3380
34	1032	1376	1721	2065	2409	2753	3097	3442
35	1042	1389	1736	2083	2430	2778	3125	3472
36	1076	1435	1794	2152	2511	2870	3228	3587
37	1102	1469	1837	2204	2571	2938	3306	3673
38	1129	1505	1882	2258	2634	3010	3387	3763
39	1152	1536	1920	2303	2687	3071	3455	3839
40	1169	1559	1949	2339	2729	3118	3508	3898

Table of Cutting Speeds
(Number Size Drills —(Continued)

Feet Per Min.	30′	40′	50′	60′	70′	80′	90′	100′
No. Size	Revolutions per Minute							
41	1194	1592	1990	2387	2785	3183	3581	3979
42	1226	1634	2043	2451	2860	3268	3677	4085
43	1288	1717	2146	2575	3004	3434	3863	4292
44	1333	1777	2221	2665	3109	3554	3999	4442
45	1397	1863	2329	2795	3261	3726	4192	4658
46	1415	1886	2358	2830	3301	3773	4244	4716
47	1460	1946	2433	2920	3406	3893	4379	4866
48	1508	2010	2513	3016	3518	4021	4523	5026
49	1570	2093	2617	3140	3663	4186	4710	5233
50	1637	2183	2729	3274	3820	4366	4911	5457
51	1710	2280	2851	3421	3991	4561	5131	5701
52	1805	2406	3008	3609	4211	4812	5414	6015
53	1924	2566	3207	3848	4490	5131	5773	6414
54	2084	2778	3473	4167	4862	5556	6251	6945
55	2204	2938	3673	4408	5142	5877	6611	7346
56	2465	3286	4108	4929	5751	6572	7394	8215
57	2671	3561	4452	5342	6232	7122	80f3	8903
58	2729	3637	4547	5456	6367	7275	8186	9095
59	2795	3726	4658	5590	6521	7453	8388	9316
60	2865	3820	4775	5729	6684	7639	8594	9549
61	2938	3918	4897	5876	6856	7835	8815	9794
62	3015	4020	5025	6030	7035	8040	9045	10050
63	3096	4128	5160	6192	7224	8256	9288	10320
64	3183	4244	5305	6366	7427	8488	9549	10610
65	3273	4364	5455	6546	7637	8728	9819	10910
66	3474	4632	5790	6948	8106	9264	10422	11580
67	3582	4776	5970	7164	8358	9552	10746	1194C
68	3696	4928	6160	7392	8624	9856	11088	12320
69	3918	5224	6530	7836	9142	10488	11754	13060
70	4091	5456	6820	8184	9548	10912	12276	13640
71	4419	5892	7365	8838	10311	11784	13257	14730
72	4584	6112	7640	9168	10696	12224	13752	15280
73	4776	6368	7960	9552	11144	12736	14328	15920
74	5106	6808	8510	10212	11914	13616	15318	17010
75	5457	7276	9095	10914	12733	14552	16371	18190
76	5730	7640	9550	11460	13370	15280	17190	19100
77	6366	8488	10610	12732	14854	16976	19098	21220
78	7161	9548	11935	14322	16709	19096	21483	23870
79	7902	10536	13170	15804	18438	21072	23706	26340
80	8490	11320	14150	16980	19810	22640	25470	28300

Table of Cutting Speeds

Letter Size Drills

Feet Per Min.	30'	40'	50'	60'	70'	80'	90'	100'
Size Letter	Revolutions per Minute							
A	491	654	818	982	1145	1309	1472	1636
B	482	642	803	963	1124	1284	1445	1605
C	473	631	789	947	1105	1262	1420	1578
D	467	622	778	934	1089	1245	1400	1556
E	458	611	764	917	1070	1222	1375	1528
F	446	594	743	892	1040	1189	1337	1486
G	440	585	732	878	1024	1170	1317	1463
H	430	574	718	862	1005	1149	1292	1436
I	421	562	702	842	983	1123	1264	1404
J	414	552	690	827	965	1103	1241	1379
K	408	544	680	815	951	1087	1223	1359
L	395	527	659	790	922	1054	1185	1317
M	389	518	648	777	907	1036	1166	1295
N	380	506	633	759	886	1012	1139	1265
O	363	484	605	725	846	967	1088	1209
P	355	473	592	710	828	946	1065	1183
Q	345	460	575	690	805	920	1035	1150
R	338	451	564	676	789	902	1014	1127
S	329	439	549	659	769	878	988	1098
T	320	426	533	640	746	853	959	1066
U	311	415	519	623	727	830	934	1038
V	304	405	507	608	709	810	912	1013
W	297	396	495	594	693	792	891	989
X	289	385	481	576	672	769	865	962
Y	284	378	473	567	662	756	851	945
Z	277	370	462	555	647	740	832	925

The range of the charts on pages 79, 80, 81, and 82 may be extended as follows. . . . To increase the cutting speed in feet per minute, increase the revolutions per minute an equal amount. Thus, for 180 fpm (which is 2 x 90 fpm) for a 7/8 diameter drill, the rpm would be 2 x 393 rpm or 786 rpm.

DECIMAL EQUIVALENTS OF NUMBER SIZE DRILLS

No.	Size in decimals	No.	Size in decimals	No.	Size in decimals	No.	Size in decimals
1	.2280	21	.1590	41	.0960	61	.0390
2	.2210	22	.1570	42	.0935	62	.0380
3	.2130	23	.1540	43	.0890	63	.0370
4	.2090	24	.1520	44	.0860	64	.0360
5	.2055	25	.1495	45	.0820	65	.0350
6	.2040	26	.1470	46	.0810	66	.0330
7	.2010	27	.1440	47	.0785	67	.0320
8	.1990	28	.1405	48	.0760	68	.0310
9	.1960	29	.1360	49	.0730	69	.0292
10	.1935	30	.1285	50	.0700	70	.0280
11	.1910	31	.1200	51	.0670	71	.0260
12	.1890	32	.1160	52	.0635	72	.0250
13	.1850	33	.1130	53	.0595	73	.0240
14	.1820	34	.1110	54	.0550	74	.0225
15	.1800	35	.1100	55	.0520	75	.0210
16	.1770	36	.1065	56	.0465	76	.0200
17	.1730	37	.1040	57	.0430	77	.0180
18	.1695	38	.1015	58	.0420	78	.0160
19	.1660	39	.0995	59	.0410	79	.0145
20	.1610	40	.0980	60	.0400	80	.0135

DECIMAL EQUIVALENTS OF LETTER SIZE DRILLS

Letter	Size in decimals	Letter	Size in decimals
A $\frac{15}{64}$	.234	N	.302
B	.238	O $\frac{5}{16}$	.316
C	.242	P $\frac{21}{64}$	.323
D	.246	Q	.332
E $\frac{1}{4}$	.250	R $\frac{11}{32}$	.339
F	.257	S	.348
G	.261	T $\frac{23}{64}$	.358
H $\frac{17}{64}$	.266	U	.368
I	.272	V $\frac{3}{8}$	.377
J	.277	W $\frac{25}{64}$	.386
K $\frac{9}{32}$	.281	X	.397
L	.290	Y $\frac{13}{32}$	.404
M $\frac{19}{64}$	.295	Z	.413

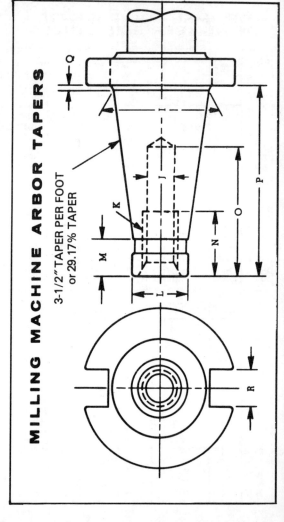

MILLING MACHINE ARBOR TAPERS

3-1/2" TAPER PER FOOT or 29.17% TAPER

84

Dimensions in Inches and *Millimeters*

Size Number	I	J	K*	L	M	N	O	P	Q	R
30	$1\frac{1}{4}$	$\frac{27}{64}$	$\frac{1}{2}$-13	0.675 / 0.673	$\frac{13}{16}$	1	2	$2\frac{3}{4}$	$\frac{1}{16}$	0.630 / 0.640
	31.75	*10.72*	*M14-2*	*17.14 / 17.09*	*20.64*	*25.40*	*50.80*	*69.85*	*1.59*	*16.00 / 16.26*
40	$1\frac{3}{4}$	$\frac{17}{32}$	$\frac{5}{8}$-11	0.987 / 0.985	1	$1\frac{1}{8}$	$2\frac{15}{16}$	$3\frac{3}{4}$	$\frac{1}{16}$	0.630 / 0.640
	44.45	*13.49*	*M16-2*	*25.07 / 25.10*	*25.40*	*28.58*	*74.61*	*95.25*	*1.59*	*16.00 / 16.26*
50	$2\frac{3}{4}$	$\frac{7}{8}$	1-8	1.549 / 1.547	$1\frac{3}{4}$	$1\frac{3}{4}$	$3\frac{1}{2}$	$5\frac{1}{8}$	$\frac{1}{8}$	1.008 / 1.018
	69.85	*22.22*	*M24-3*	*39.34 / 39.29*	*25.40*	*44.45*	*88.90*	*130.18*	*3.18*	*25.86 / 25.60*
60	$4\frac{1}{4}$	$1\frac{7}{64}$	$1\frac{1}{4}$-7	2.361 / 2.359	$1\frac{3}{4}$	$2\frac{1}{4}$	$4\frac{1}{4}$	$8\frac{5}{16}$	$\frac{1}{8}$	1.008 / 1.018
	107.95	*28.12*	*M27-3*	*59.97 / 59.92*	*44.45*	*57.15*	*107.95*	*211.14*	*3.18*	*25.86 / 25.60*

*Standards are not established for the metric thread sizes.

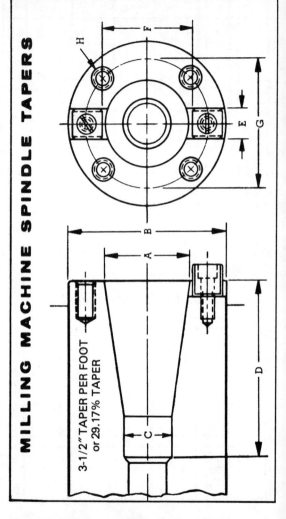

MILLING MACHINE SPINDLE TAPERS

3-1/2" TAPER PER FOOT or 29.17% TAPER

Dimensions in Inches and *Millimeters*

Size Number	A	B	C	D	E	F	G	H*
30	1¼	2.7493 / 2.7488	0.692 / 0.685	2⅞	0.6255 / 0.6252	1.315 / 1.285	2.130 / 2.120	⅜-16
	31.75	*69.83 / 69.82*	*17.58 / 17.40*	*73.02*	*15.89 / 15.88*	*33.40 / 32.64*	*54.10 / 53.85*	*M10 - 1.5*
40	1¾	3.4993 / 3.4988	1.005 / 0.997	3⅞	0.6255 / 0.6252	1.819 / 1.807	2.630 / 2.620	½-13
	44.45	*88.88 / 88.87*	*25.53 / 25.32*	*94.42*	*15.89 / 15.88*	*46.20 / 45.90*	*66.80 / 66.55*	*M14 - 2*
50	2¾	5.0618 / 5.0613	1.568 / 1.559	5½	1.0006 / 1.0002	2.819 / 2.807	4.005 / 3.995	⅝-11
	69.85	*128.57 / 128.56*	*39.83 / 39.60*	*139.70*	*25.42 / 25.40*	*71.60 / 71.30*	*101.73 / 101.47*	*M16 - 2*
60	4¼	8.7180 / 8.7175	2.381 / 2.371	8⅝	1.000 / 0.999	4.819 / 4.807	7.005 / 6.995	¾-10
	107.95	*221.44 / 221.42*	*60.48 / 60.22*	*219.07*	*25.40 / 25.38*	*122.40 / 122.10*	*177.93 / 177.67*	*M20 - 2.5*

* Standards are not established for the metric thread sizes.

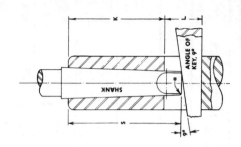

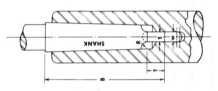

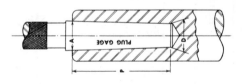

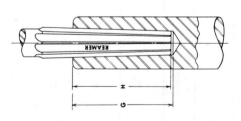

American National Standard Tapers

Detail Dimensions†

American National Standard Taper Number	Diam. of Plug at Small End	Diam. at End of Socket	Shank: Whole Length	Shank: Depth	Depth of Drilled Hole	Depth of Reamed Hole	Standard Plug Depth	Tang: Thickness	Tang: Length	Tang: Radius	Tang: Radius	Tang Slot: Width	Tang Slot: Length	End of Socket to Tang Slot	Taper per Inch	Taper per Foot	American National Standard Taper Number
	D	A	B	S	G	H	P	t	T	R	a	W	L	K			
‡0	.25200	.35610	2 11/32	2 7/32	2 1/16	2 1/32	2	5/32	1/4	5/32	3/64	11/64	9/16	1 15/16	.052050	.62460	‡0
1	.36900	.47500	2 9/16	2 7/16	2 3/16	2 5/32	2 1/8	13/64	3/8	3/16	3/64	7/32	3/4	2 1/16	.049882	.59858	1
2	.57200	.70000	3 1/8	2 15/16	2 21/32	2 39/64	2 9/16	1/4	7/16	1/4	1/16	17/64	7/8	2 1/2	.049951	.59941	2
3	.77800	.93800	3 7/8	3 11/16	3 5/16	3 1/4	3 3/16	5/16	9/16	9/32	5/64	21/64	1 3/16	3 1/16	.050196	.60235	3
4	1.02000	1.23100	4 7/8	4 5/8	4 3/16	4 1/8	4 1/16	15/32	5/8	5/16	3/32	31/64	1 1/4	3 7/8	.051938	.62326	4
4 1/2	1.26600	1.50000	5 3/8	5 1/8	4 5/8	4 9/16	4 1/2	9/16	11/16	3/8	1/8	37/64	1 3/8	4 5/16	.052000	.62400	4 1/2
5	1.47500	1.74800	6 1/8	5 7/8	5 5/16	5 1/4	5 3/16	5/8	3/4	3/8	1/8	21/32	1 1/2	4 15/16	.052626	.63151	5
6	2.11600	2.49400	8 9/16	8 1/4	7 13/32	7 21/64	7 1/4	3/4	1 1/8	1/2	5/32	25/32	1 3/4	7	.052138	.62565	6
7	2.75000	3.27000	11 5/8	11 1/4	10 5/32	10 5/64	10	1 1/8	1 3/8	3/4	3/16	1 5/32	2 5/8	9 1/2	.052000	.62400	7

†Table agrees with American National Standards for Taper Shanks except for angle and undercut of tang.

‡Size 0 taper shank not listed in American National Standards.

Brown & Sharpe Taper Shanks

Sec. III
Tapers

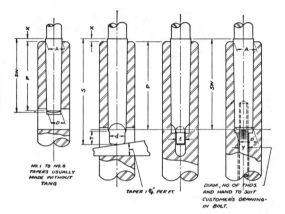

NO. 1 TO NO. 6 TAPERS USUALLY MADE WITHOUT TANG

TAPER 1¾ PER FT.

DIAM., NO. OF THDS. AND HAND TO SUIT CUSTOMER'S DRAWING—IN BOLT.

No. of Taper	Diam. of Plug at Small End D	Diam. at End of Socket A	Plug Depth P	Shank Length (with Tang) S	Shank Length (without Tang) SW	Amount Shank Projects from End of Socket X	Width of Tongue t	Diam. of Tongue d	Length of Tongue T	Taper per Foot
1	.2000	.2392	15/16	1 9/32	1 1/16	3/32	1/8	.170	3/16	.50200
2	.2500	.2997	1 3/16	1 19/32	1 11/32	3/32	5/32	.220	1/4	.50200
3	.3120	.3752	1 1/2	1 31/32	1 11/32	3/32	3/16	.282	5/16	.50200
4	.3500	.4023	1 1/4	1 3/4	1 25/64	3/32	7/32	.320	11/32	.50240
5	.4500	.5231	1 3/4	2 9/32	1 29/32	3/32	1/4	.420	3/8	.50160
6	.5000	.5996	2 3/8	2 31/32	2 17/32	3/32	9/32	.460	7/16	.50329
7	.6000	.7254	3	3 5/8	3 3/32	3/32	5/16	.560	15/32	.50147
8	.7500	.8987	3 9/16	4 1/4	3 11/16	1/8	11/32	.710	1/2	.50100
9	.9001	1.0670	4	4 3/4	4 1/8	1/8	3/8	.860	9/16	.50085
10	1.0446	1.2892	5 11/16	6 17/32	5 13/16	1/8	7/16	1.010	21/32	.51612
11	1.2500	1.5318	6 3/4	7 19/32	6 7/8	1/8	7/16	1.210	21/32	.50100
12	1.5001	1.7968	7 1/8	8 1/16	7 1/4	1/8	1/2	1.460	3/4	.49973
13	1.7500	2.0730	7 3/4	8 11/16	7 7/8	1/8	1/2	1.710	3/4	.50020
14	2.0000	2.3437	8 1/4	9 9/32	8 3/8	1/8	9/16	1.960	27/32	.50000
15	2.2500	2.6146	8 3/4	9 25/32	8 7/8	1/8	9/16	2.210	27/32	.50000
16	2.5000	2.8854	9 1/4	10 3/8	9 3/8	1/8	5/8	2.450	15/16	.50000
17	2.7500	3.1562	9 3/4	....	9 7/8	1 1/8	....	....	....	.50000
18	3.0000	3.4271	10 1/4	....	10 3/8	1/8	....	....	....	.50000

Rules for Figuring Tapers

Given	To Find	Rule
The taper per foot.......	The taper per inch.......	Divide the taper per foot by 12.
The taper per inch.......	The taper per foot.......	Multiply the taper per inch by 12.
End diameters and length of taper in inches.......	The taper per foot.......	Subtract small diameter from large; divide by length of taper, and multiply quotient by 12.
Large diameter and length of taper in inches and taper per foot.........	Diameter at small end in inches..................	Divide taper per foot by 12; multiply by length of taper, and subtract result from large diameter.
Small diameter and length of taper in inches, and taper per foot........	Diameter at large end in inches..................	Divide taper per foot by 12; multiply by length of taper; and add result to small diameter.
The taper per foot and two diameters in inches......	Distance between two given diameters in inches......	Subtract small diameter from large; divide remainder by taper per foot, and multiply quotient by 12.
The taper per foot........	Amount of taper in a certain length given in inches....	Divide taper per foot by 12; multiply by given length of tapered part.

TAPERS

Tapers From 1/16 to 1 1/4 Inch Per Foot

Amount of Taper for Lengths up to 24 Inches

Length Tapered Portion Inches	Taper Per Foot									
	1/16	3/32	1/8	1/4	3/8	1/2	5/8	3/4	1	1 1/4
1/32	.0002	.0002	.0003	.0007	.0010	.0013	.0016	.0020	.0026	.0033
1/16	.0003	.0005	.0007	.0013	.0020	.0026	.0033	.0039	.0052	.0065
1/8	.0007	.0010	.0013	.0026	.0039	.0052	.0065	.0078	.0104	.0130
3/16	.0010	.0015	.0020	.0039	.0059	.0078	.0098	.0117	.0156	.0195
1/4	.0013	.0020	.0026	.0052	.0078	.0104	.0130	.0156	.0208	.0260
5/16	.0016	.0024	.0033	.0065	.0098	.0130	.0163	.0195	.0260	.0326
3/8	.0020	.0029	.0039	.0078	.0117	.0156	.0195	.0234	.0312	.0391
7/16	.0023	.0034	.0046	.0091	.0137	.0182	.0228	.0273	.0365	.0456
1/2	.0026	.0039	.0052	.0104	.0156	.0208	.0260	.0312	.0417	.0521
9/16	.0029	.0044	.0059	.0117	.0176	.0234	.0293	.0352	.0469	.0586
5/8	.0033	.0049	.0065	.0130	.0195	.0260	.0326	.0391	.0521	.0651
11/16	.0036	.0054	.0072	.0143	.0215	.0286	.0358	.0430	.0573	.0716
3/4	.0039	.0059	.0078	.0156	.0234	.0312	.0391	.0469	.0625	.0781
13/16	.0042	.0063	.0085	.0169	.0254	.0339	.0423	.0508	.0677	.0846
7/8	.0046	.0068	.0091	.0182	.0273	.0365	.0456	.0547	.0729	.0911
15/16	.0049	.0073	.0098	.0195	.0293	.0391	.0488	.0586	.0781	.0977
1	.0052	.0078	.0104	.0208	.0312	.0417	.0521	.0625	.0833	.1042
2	.0104	.0156	.0208	.0417	.0625	.0833	.1042	.1250	.1667	.2083
3	.0156	.0234	.0312	.0625	.0937	.1250	.1562	.1875	.2500	.3125
4	.0208	.0312	.0417	.0833	.1250	.1667	.2083	.2500	.3333	.4167
5	.0260	.0391	.0521	.1042	.1562	.2083	.2604	.3125	.4167	.5208
6	.0312	.0469	.0625	.1250	.1875	.2500	.3125	.3750	.5000	.6250
7	.0365	.0547	.0729	.1458	.2187	.2917	.3646	.4375	.5833	.7292
8	.0417	.0625	.0833	.1667	.2500	.3333	.4167	.5000	.6667	.8333
9	.0469	.0703	.0937	.1875	.2812	.3750	.4687	.5625	.7500	.9375
10	.0521	.0781	.1042	.2083	.3125	.4167	.5208	.6250	.8333	1.0417
11	.0573	.0859	.1146	.2292	.3437	.4583	.5729	.6875	.9167	1.1458
12	.0625	.0937	.1250	.2500	.3750	.5000	.6250	.7500	1.0000	1.2500
13	.0677	.1016	.1354	.2708	.4062	.5417	.6771	.8125	1.0833	1.3542
14	.0729	.1094	.1458	.2917	.4375	.5833	.7292	.8750	1.1667	1.4583
15	.0781	.1172	.1562	.3125	.4687	.6250	.7812	.9375	1.2500	1.5625
16	.0833	.1250	.1667	.3333	.5000	.6667	.8333	1.0000	1.3333	1.6667
17	.0885	.1328	.1771	.3542	.5312	.7083	.8854	1.0625	1.4167	1.7708
18	.0937	.1406	.1875	.3750	.5625	.7500	.9375	1.1250	1.5000	1.8750
19	.0990	.1484	.1979	.3958	.5937	.7917	.9896	1.1875	1.5833	1.9792
20	.1042	.1562	.2083	.4167	.6250	.8333	1.0417	1.2500	1.6667	2.0833
21	.1094	.1641	.2187	.4375	.6562	.8750	1.0937	1.3125	1.7500	2.1875
22	.1146	.1719	.2292	.4583	.6875	.9167	1.1458	1.3750	1.8333	2.2917
23	.1198	.1797	.2396	.4792	.7187	.9583	1.1979	1.4375	1.9167	2.3958
24	.1250	.1875	.2500	.5000	.7500	1.0000	1.2500	1.5000	2.0000	2.5000

Rules and Formulas for Figuring Tapers

(Applies to both inch and millimeter dimensions)

Notation: D = Large diameter; d = Small diameter; L = Distance between diameters; % = Percentage of taper.

Given	To Find	Rules	Formulas
End diameters and length of taper	Percentage of taper (%)	Multiply the difference in diameters by 100; divide this result by the length of taper.	$\% = \dfrac{(D - d) \times 100}{L}$
Small diameter and length of taper	Diameter at the large end (D)	Multiply the distance between diameters by the percentage of taper, and divide by 100; add this result to the small diameter.	$D = d + \dfrac{L \times \%}{100}$
Large diameter and length of taper	Diameter at the small end (d)	Multiply the distance between diameters by the percentage of taper, and divide by 100; subtract this result from the large diameter.	$d = D - \dfrac{L \times \%}{100}$
Large and small diameters and percentage of taper	Distance between the large and small diameters (L)	Multiply the difference in diameters by 100; divide this result by the percentage of taper.	$L = \dfrac{(D - d) \times 100}{\%}$
The distance between the two diameters and the percentage of taper	The difference in diameters (D − d)	Multiply the distance between diameters by the percentage of taper; divide this result by 100.	$(D - d) = \dfrac{L \times \%}{100}$
Percentage of taper	Tangent of the single angle of the taper	Divide the percentage of taper by 200.	$\text{Tangent} = \dfrac{\%}{200}$

AMOUNT OF TAPER FOR LENGTHS UP TO 600 MILLIMETERS

Tapers from 1% to 30%

Length of taper in mm	Percentage of taper									
	1%	2%	3%	4%	5%	10%	15%	20%	25%	30%
1	.01	.02	.03	.04	.05	.10	.15	.20	.25	.30
2	.02	.04	.06	.08	.10	.20	.30	.40	.50	.60
3	.03	.06	.09	.12	.15	.30	.45	.60	.75	.90
4	.04	.08	.12	.16	.20	.40	.60	.80	1.00	1.20
5	.05	.10	.15	.20	.25	.50	.75	1.00	1.25	1.50
6	.06	.12	.18	.24	.30	.60	.90	1.20	1.50	1.80
7	.07	.14	.21	.28	.35	.70	1.05	1.40	1.75	2.10
8	.08	.16	.24	.32	.40	.80	1.20	1.60	2.00	2.40
9	.09	.18	.27	.36	.45	.90	1.35	1.80	2.25	2.70
10	.10	.20	.30	.40	.50	1.00	1.50	2.00	2.50	3.00
20	.20	.40	.60	.80	1.00	2.00	3.00	4.00	5.00	6.00
30	.30	.60	.90	1.20	1.50	3.00	4.50	6.00	7.50	9.00
40	.40	.80	1.20	1.60	2.00	4.00	6.00	8.00	10.00	12.00
50	.50	1.00	1.50	2.00	2.50	5.00	7.50	10.00	12.50	15.00
60	.60	1.20	1.80	2.40	3.00	6.00	9.00	12.00	15.00	18.00
70	.70	1.40	2.10	2.80	3.50	7.00	10.50	14.00	17.50	21.00
80	.80	1.60	2.40	3.20	4.00	8.00	12.00	16.00	20.00	24.00
90	.90	1.80	2.70	3.60	4.50	9.00	13.50	18.00	22.50	27.00
100	1.00	2.00	3.00	4.00	5.00	10.00	15.00	20.00	25.00	30.00
200	2.00	4.00	6.00	8.00	10.00	20.00	30.00	40.00	50.00	60.00
300	3.00	6.00	9.00	12.00	15.00	30.00	45.00	60.00	75.00	90.00
400	4.00	8.00	12.00	16.00	20.00	40.00	60.00	80.00	100.00	120.00
500	5.00	10.00	15.00	20.00	25.00	50.00	75.00	100.00	125.00	150.00
600	6.00	12.00	18.00	24.00	30.00	60.00	90.00	120.00	150.00	180.00

TAPERS PER FOOT

AND

CORRESPONDING ANGLES

Taper per Foot	Included Angle	Angle with Center Line	Taper per Foot	Included Angle	Angle with Center Line
$\frac{1}{16}$	0° 17′ 53″	0° 8′ 57″	2¼	10° 42′ 41″	5° 21′ 21″
⅛	0 35 47	0 17 54	2⅜	11 18 12	5 39 6
$\frac{3}{16}$	0 53 44	0 26 52	2½	11 53 38	5 56 49
¼	1 11 38	0 35 49	2⅝	12 29 2	6 14 31
$\frac{5}{16}$	1 29 31	0 44 46	2¾	13 4 25	6 32 13
⅜	1 47 25	0 53 42	2⅞	13 39 44	6 49 52
$\frac{7}{16}$	2 5 18	1 2 39	3	14 15 0	7 7 30
½	2 23 12	1 11 36	3¼	15 25 27	7 42 43
$\frac{9}{16}$	2 41 7	1 20 34	3½	16 35 41	8 17 50
⅝	2 59 3	1 29 31	3¾	17 45 40	8 52 50
$\frac{11}{16}$	3 16 56	1 38 28	4	18 55 31	9 27 45
¾	3 34 48	1 47 24	4¼	20 5 1	10 2 31
$\frac{13}{16}$	3 52 42	1 56 21	4½	21 14 20	10 37 10
⅞	4 10 32	2 5 16	4¾	22 23 27	11 11 43
$\frac{15}{16}$	4 28 26	2 14 13	5	23 32 12	11 46 6
1	4 46 19	2 23 10	5¼	24 40 43	12 20 21
1⅛	5 22 2	2 41 1	5½	25 48 53	12 54 27
1¼	5 57 45	2 58 53	5¾	26 56 48	13 28 24
1⅜	6 33 29	3 16 44	6	28 4 20	14 2 10
1½	7 9 10	3 34 35	6¼	29 11 36	14 35 48
1⅝	7 44 49	3 52 24	6½	30 18 28	15 9 14
1¾	8 20 28	4 10 14	6¾	31 25 2	15 42 31
1⅞	8 56 2	4 28 1	7	32 31 14	16 15 37
2	9 31 37	4 45 49	7½	34 42 30	17 21 15
2⅛	10 7 11	5 3 35	8	35 47 32	17 53 46

TO FIND INCLUDED ANGLE FOR GIVEN TAPER PER FOOT. When the taper in inches per foot is known, and the corresponding included angle is required, divide the taper in inches per foot by 24; find the angle corresponding to this quotient in a table of tangents, and double this angle.

PERCENTAGE OF TAPERS
AND
CORRESPONDING ANGLES

Percentage of Taper	Angle with Center Line	Included Angle	Percentage of Taper	Angle with Center Line	Included Angle
1	0° 17′ 11″	0° 34′ 23″	26	7° 24′ 25″	14° 48′ 50″
2	0 34 23	1 8 45	27	7 41 18	15 22 37
3	0 51 34	1 43 8	28	7 58 11	15 48 21
4	1 8 45	2 17 29	29	8 15 1	16 30 3
5	1 25 56	2 51 51	30	8 31 51	17 3 42
6	1 43 6	3 26 12	31	8 48 39	17 37 17
7	2 0 16	4 0 33	32	9 5 25	18 10 50
8	2 17 26	4 34 53	33	9 22 10	18 44 20
9	2 34 36	5 9 12	34	9 38 53	19 17 46
10	2 51 45	5 43 30	35	9 55 35	19 51 9
11	3 8 53	6 17 46	36	10 12 14	20 24 29
12	3 26 1	6 52 2	37	10 28 52	20 57 45
13	3 43 8	7 23 17	38	10 45 29	21 30 58
14	4 0 15	8 0 30	39	11 2 3	22 4 6
15	4 17 21	8 34 42	40	11 18 36	22 37 12
16	4 34 26	9 8 52	41	11 35 7	23 10 13
17	4 51 31	9 43 1	42	11 51 35	23 43 11
18	5 8 34	10 17 8	43	12 8 2	24 16 4
19	5 25 37	10 51 13	44	12 24 27	24 48 54
20	5 42 38	11 25 16	45	12 40 49	25 21 39
21	5 59 39	11 59 18	46	12 57 10	25 54 20
22	6 16 38	12 33 17	47	13 13 29	26 26 57
23	6 33 37	13 7 14	48	13 29 45	26 59 30
24	6 50 34	13 41 8	49	13 45 59	27 31 58
25	7 7 30	14 15 0	50	14 2 10	28 4 21

TO FIND THE INCLUDED ANGLE FOR A GIVEN PERCENTAGE OF TAPER. When the percentage of taper is known, and the corresponding included angle is required, divide the percentage of taper by 200; find the angle corresponding to this quotient in a table of tangents, and double this angle.

Note: A one percent taper is equal to .010″ taper in 1.000″, or 1 millimeter taper in 100 millimeters.

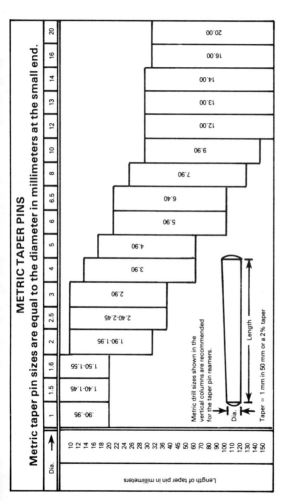

METRIC TAPER PINS

Metric taper pin sizes are equal to the diameter in millimeters at the small end.

Dia. →	1	1.5	1.6	2	2.5	3	4	5	6	6.5	8	10	12	13	14	16	20
	.90-.95	1.40-1.45	1.50-1.55	1.90-1.95	2.40-2.45	2.90	3.90	4.90	5.90	6.40	7.90	9.90	12.00	13.00	14.00	16.00	20.00

Length of taper pin in millimeters

10
12
14
16
18
20
22
24
26
28
30
32
36
40
45
50
60
70
80
90
100
110
120
130
140
150

Metric drill sizes shown in the vertical columns are recommended for the taper pin reamers.

Dia.

Length

Taper = 1 mm in 50 mm or a 2% taper

Diameter of Taper Pin at Small End

and Recommended Drill Sizes for Taper Pin Reamers

Length Pin \ No. of Pin	2/0	0	1	2	3	4	5	6	7	8	9	10
1/2"	.1306 No. 30	.1456 No. 27	.1616 No. 21	.1826 No. 15								
5/8"	.128 1/8"	.143 No. 28	.159 No. 22	.180 No. 16								
3/4"	.1254 No. 31	.1404 No. 29	.1564 No. 23	.1774 No. 17	.2034 13/64"	.2344 A	.2734 I	.3254 P				
7/8"	.1228 No. 31	.1378 No. 29	.1538 No. 24	.1748 No. 17	.2008 No. 8	.2318 No. 1	.2708 H	.3228 O				
1"	.1202 No. 32	.1352 No. 30	.1512 No. 25	.1722 No. 18	.1982 No. 9	.2292 No. 1	.2682 H	.3202 O	.3882 W			
1 1/8"	.1176 No. 33	.1326 No. 30	.1486 No. 26	.1696 No. 18	.1956 No. 10	.2266 No. 2	.2656 G	.3176 O	.3856 V			
1 1/4"	.115 No. 33	.130 No. 30	.146 No. 27	.167 No. 19	.193 No. 11	.224 No. 2	.263 G	.315 5/16"	.383 V	.466 29/64"		
1 1/2"				.1618 No. 21	.1878 No. 13	.2188 No. 3	.2578 1/4"	.3098 N	.3778 3/8"	.4608 29/64"	.5598 35/64"	.6748 43/64"
1 5/8"					.1852 No. 14	.2162 No. 3	.2552 1/4"	.3072 N	.3752 U	.4582 29/64"	.5572 35/64"	.6722 43/64"
1 3/4"						.2136 No. 3	.2526 1/4"	.3046 N	.3726 U	.4556 29/64"	.5546 35/64"	.6696 21/32"
1 7/8"							.2495 D	.3015 19/64"	.3695 U	.4525 7/16"	.5515 35/64"	.6665 21/32"

Size					
2"	.2994 $^{19/64"}$	.3674 $^{23/64"}$	.4504 $^{7/16"}$	.5494 $^{35/64"}$	.6644 $^{29/32"}$
2¼"	.2942 L	.3622 $^{23/64"}$	.4452 $^{7/16"}$	.5442 $^{17/32"}$	.6592 $^{21/32"}$
2½"	.289 $^{9/32"}$	.357 S	.440 $^{7/16"}$	.539 $^{17/32"}$	.654 $^{41/64"}$
2¾"		.3518 S	.4348 $^{27/64"}$	.5338 $^{17/32"}$	.6488 $^{41/64"}$
3"		.3466 $^{11/32"}$	.4296 $^{27/64"}$	.5286 $^{33/64"}$	.6436 $^{41/64"}$
3¼"		.3414 R	.4244 $^{27/64"}$	.5234 $^{33/64"}$	.6384 $^{5/8"}$
3½"			.4192 Z	.5182 $^{33/64"}$	.6332 $^{5/8"}$
3¾"			.414 Z	.513 $^{1/2"}$	.628 $^{5/8"}$
4"			.4088 $^{13/32"}$	.5078 $^{1/2"}$	.6228 $^{9/64"}$
4¼"				.5026 $^{1/2"}$	.6176 $^{19/64"}$
4½"				.4974 $^{31/64"}$	.6124 $^{19/64"}$
4¾"				.4922 $^{31/64"}$	.6072 $^{19/32"}$
5"					.602 $^{19/32"}$
5¼"					.5968 $^{19/32"}$
5½"					.5916 $^{37/64"}$

Sec. III
Tapers

Basic Thread Dimensions and Tap Drill Sizes

American National Coarse Thread

Formerly A. S. M. E. Regular for Sizes 1-12; U. S. Standard for Sizes ¼″ and larger

D = Major Dia.
E = Pitch Dia.
K = Minor Dia.

Size of Thread and Threads per Inch	Major Diameter D Inches	Pitch Diameter E Inches	Minor Diameter Internal Threads K Inches	Commercial Tap Drill to Produce Approx. 75% Full Thread	Decimal Equivalent of Tap Drill Inches
1 x64	.0730	.0629	.0561	No. 53	.0595
2 x56	.0860	.0744	.0667	No. 50	.0700
3 x48	.0990	.0855	.0764	No. 47	.0785
4 x40	.1120	.0958	.0849	No. 43	.0890
5 x40	.1250	.1088	.0979	No. 38	.1015
6 x32	.1380	.1177	.1042	No. 36	.1065
8 x32	.1640	.1437	.1302	No. 29	.1360
10 x24	.1900	.1629	.1449	No. 25	.1495
12 x24	.2160	.1889	.1709	No. 16	.1770
¼x20	.2500	.2175	.1959	No. 7	.2010
5⁄16x18	.3125	.2764	.2524	F	.2570
3⁄8x16	.3750	.3344	.3073	5⁄16	.3125
7⁄16x14	.4375	.3911	.3602	U	.3680
½x13	.5000	.4500	.4167	27⁄64	.4219
9⁄16x12	.5625	.5084	.4723	31⁄64	.4844
5⁄8x11	.6250	.5660	.5266	17⁄32	.5312
¾x10	.7500	.6850	.6417	21⁄32	.6562
7⁄8x 9	.8750	.8028	.7547	49⁄64	.7656
1 x 8	1.0000	.9188	.8647	7⁄8	.8750
1⅛x 7	1.1250	1.0322	.9704	63⁄64	.9844
1¼x 7	1.2500	1.1572	1.0954	1 7⁄64	1.1094
1⅜x 6	1.3750	1.2667	1.1946	1 7⁄32	1.2188
1½x 6	1.5000	1.3917	1.3196	1 11⁄32	1.3438
1¾x 5	1.7500	1.6201	1.5335	1 9⁄16	1.5625
2 x 4½	2.0000	1.8557	1.7594	1 25⁄32	1.7812

Basic Thread Dimensions and Tap Drill Sizes

American National Fine Threads

Formerly A. S. M. E. Special for Sizes 0-12; S. A. E. Standard for Sizes ¼ in. and larger

Size of Thread and Threads per inch	Major Diameter D Inches	Pitch Diameter E Inches	Minor Diameter Internal Threads Kn Inches	Commercial Tap Drill to Produce Approx. 75% Full Thread	Decimal Equivalent of Tap Drill Inches
0 x80	.0600	.0519	.0465	³⁄₆₄	.0469
1 x72	.0730	.0640	.0580	¹⁄₁₆	.0625
2 x64	.0860	.0759	.0691	No. 49	.0730
3 x56	.0990	.0874	.0797	No. 44	.0860
4 x48	.1120	.0985	.0894	No. 42	.0935
5 x44	.1250	.1102	1004	No. 36	.1065
6 x40	.1380	.1218	.1109	No. 32	.1160
8 x36	.1640	.1460	.1339	Nc. 29	.1360
10 x32	.1900	.1697	.1562	No. 20	.1610
12 x28	.2160	.1928	.1773	No. 14	.1820
¼ x28	.2500	.2268	.2113	⁷⁄₃₂	.2188
⁵⁄₁₆ x24	.3125	.2854	.2674	I	.2720
⅜ x24	.3750	.3479	.3299	Q	.3320
⁷⁄₁₆ x20	.4375	.4050	.3834	²⁵⁄₆₄	.3906
½ x20	.5000	.4675	.4459	²⁹⁄₆₄	.4531
⁹⁄₁₆ x18	.5625	.5264	.5024	³³⁄₆₄	.5156
⅝ x18	.6250	.5889	.5649	³⁷⁄₆₄	.5781
¾ x16	.7500	.7094	.6823	¹¹⁄₁₆	.6875
⅞ x14	.8750	.8286	.7977	¹³⁄₁₆	.8125
1 x12	1.0000	.9459	.9098	⁵⁹⁄₆₄	.9219
1⅛ x12	1.1250	1.0709	1.0348	1³⁄₆₄	1.0469
1¼ x12	1.2500	1.1959	1.1598	1¹¹⁄₆₄	1.1719
1⅜ x12	1.3750	1.3209	1.2848	1¹⁹⁄₆₄	1.2969
1½ x12	1.5000	1.4459	1.4098	1²⁷⁄₆₄	1.4219

Basic Thread Dimensions and Tap Drill Sizes
American National Form Thread Special Pitches

Size of Thread and Threads per Inch	Major Diameter D Inches	Pitch Diameter E Inches	Minor Diameter K Inches	Commercial Tap Drill to Produce Approx. 75% Full Thread	Decimal Equivalent of Tap Drill Inches
¼ x24	.2500	.2229	.2049	No. 4	.2090
¼ x32	.2500	.2297	.2162	⁷⁄₃₂	.2288
⁵⁄₁₆ x20	.3125	.2800	.2584	¹⁷⁄₆₄	.2656
⁵⁄₁₆ x32	.3125	.2922	.2787	⁹⁄₃₂	.2812
⅜ x20	.3750	.3425	.3209	²¹⁄₆₄	.3281
⅜ x32	.3750	.3547	.3412	¹¹⁄₃₂	.3438
⁷⁄₁₆ x24	.4375	.4104	.3924	X	.3970
⁷⁄₁₆ x28	.4375	.4143	.3988	Y	.4040
½ x24	.5000	.4729	.4549	²⁹⁄₆₄	.4531
½ x28	.5000	.4768	.4613	¹⁵⁄₃₂	.4688
⁹⁄₁₆ x24	.5625	.5354	.5174	³³⁄₆₄	.5156
⅝ x12	.6250	.5709	.5348	³⁵⁄₆₄	.5469
⅝ x24	.6250	.5979	.5799	³⁷⁄₆₄	.5781
¹¹⁄₁₆x12	.6875	.6334	.5973	³⁹⁄₆₄	.6094
¹¹⁄₁₆x24	.6875	.6604	.6424	⁴¹⁄₆₄	.6406
¾ x12	.7500	.6959	.6598	⁴³⁄₆₄	.6719
¾ x20	.7500	.7175	.6959	⁴⁵⁄₆₄	.7031
¹³⁄₁₆x12	.8125	.7584	.7223	⁴⁷⁄₆₄	.7344
¹³⁄₁₆x16	.8125	.7719	.7448	¾	.7500
¹³⁄₁₆x20	.8125	.7800	.7584	⁴⁹⁄₆₄	.7656
⅞ x12	.8750	.8209	.7848	⁵¹⁄₆₄	.7969
⅞ x16	.8750	.8344	.8073	¹³⁄₁₆	.8125
⅞ x20	.8750	.8425	.8209	⁵³⁄₆₄	.8281
¹⁵⁄₁₆x12	.9375	.8834	.8473	⁵⁵⁄₆₄	.8594
¹⁵⁄₁₆x16	.9375	.8969	.8698	⅞	.8750
¹⁵⁄₁₆x20	.9375	.9050	.8834	⁵⁷⁄₆₄	.8906
1 x14	1.0000	.9536	.9227	¹⁵⁄₁₆	.9375
1 x16	1.0000	.9594	.9323	¹⁵⁄₁₆	.9375
1 x20	1.0000	.9675	.9459	⁶¹⁄₆₄	.9531

BASIC THREAD SHAPES

Notation and Formulas

F = WIDTH OF FLAT AT TOP   P = PITCH = $\frac{1}{N}$ ON ALL THREADS

D = DEPTH OF THREAD   N = NO. OF THREADS PER INCH

F' = WIDTH OF FLAT AT BOTTOM OF THREAD

SHARP V (THEORETICAL) THREAD

D = DEPTH = .86603 × P

AMERICAN NATIONAL THREAD

D = DEPTH = .64952 × P   F = FLAT = $\frac{P}{8}$

ACME THREAD

F = .3707 × P   D = $\frac{P}{2}$ + .010

F' = .3707 × P − .0052

WORM THREAD

D = .6866 × P

SQUARE THREAD

BUTTRESS THREAD

103

Constants for Finding Pitch Diameter

and

Root Diameter of Screw Threads

To find the pitch diameter or root diameter of any screw thread, substract the constant for the number of threads per inch from the outside diameter.

Threads per Inch	Constants for Finding Pitch Diameter			Constants for Finding Root Diameter		
	National Thread	Whitworth Thread	Theoretical V	National Thread	Whitworth Thread	Theoretical V
72	.00902	.00889	.01203	.01804	.01786	.02406
64	.01015	.01000	.01353	.02030	.02001	.02706
60	.01083	.01067	.01443	.02165	.02134	.02887
56	.01160	.01144	.01546	.02320	.02286	.03093
50	.01299	.01281	.01732	.02598	.02562	.03464
48	.01353	.01334	.01804	.02706	.02668	.03608
44	.01476	.01455	.01968	.02952	.02910	.03936
40	.01624	.01601	.02165	.03248	.03202	.04330
36	.01804	.01779	.02406	.03608	.03558	.04811
32	.02030	.02001	.02706	.04059	.04002	.05413
30	.02165	.02134	.02887	.04330	.04268	.05773
28	.02320	.02287	.03093	.04639	.04574	.06186
27	.02406	.02372	.03207	.04812	.04742	.06416
26	.02498	.02463	.03331	.04996	.04926	.06662
24	.02706	.02668	.03608	.05413	.05336	.07217
22	.02952	.02911	.03936	.05905	.05821	.07873
20	.03248	.03202	.04330	.06495	.06403	.08660
18	.03608	.03557	.04811	.07217	.07114	.09623
16	.04059	.04002	.05413	.08119	.08004	.10825
14	.04639	.04574	.06186	.09279	.09147	.12372
13	.04996	.04926	.06662	.09993	.09851	.13323
12	.05413	.05336	.07217	.10825	.10672	.14434
11½	.05648	.05568	.07531	.11296	.11132	.15062
11	.05905	.05821	.07873	.11809	.11642	.15746
10	.06495	.06403	.08660	.12990	.12806	.17321
9	.07217	.07115	.09623	.14434	.14230	.19245
8	.08119	.08004	.10825	.16238	.16008	.21651
7	.09279	.09148	.12372	.18558	.18295	.24744
6	.10825	.10672	.14434	.21651	.21344	.28868
5½	.11809	.11642	.15746	.23619	.23284	.31492
5	.12990	.12807	.17321	.25981	.25613	.34641
4½	.14434	.14230	.19245	.28868	.28458	.38490
4	.16238	.16008	.21651	.32476	.32017	.43301
3½	.18558	.18295	.24744	.37115	.36590	.49487
3¼	.19985	.19702	.26647	.39970	.39404	.53294
3	.21651	.21344	.28868	.43301	.42689	.57733

National Screw Thread Standards

Class II, Screws and Nuts
National Coarse Series
Pitch Diameter Limits

Size	Threads per Inch	Basic Pitch Diameter	Screws		Nuts	
			Minimum	Maximum	Minimum	Maximum
1	64	.0629	.0610	.0629	.0629	.0648
2	56	.0744	.0724	.0744	.0744	.0764
3	48	.0855	.0833	.0855	.0855	.0877
4	40	.0958	.0934	.0958	.0958	.0982
5	40	.1088	.1064	.1088	.1088	.1112
6	32	.1177	.1150	.1177	.1177	.1204
8	32	.1437	.1410	.1437	.1437	.1464
10	24	.1629	.1596	.1629	.1629	.1662
12	24	.1889	.1856	.1889	.1889	.1922
¼	20	.2175	.2139	.2175	.2175	.2211
5⁄16	18	.2764	.2723	.2764	.2764	.2805
⅜	16	.3344	.3299	.3344	.3344	.3389
7⁄16	14	.3911	.3862	.3911	.3911	.3960
½	13	.4500	.4448	.4500	.4500·	.4552
9⁄16	12	.5084	.5028	.5084	.5084	.5140
⅝	11	.5660	.5601	.5660	.5660	.5719
¾	10	.6850	.6786	.6850	.6850	.6914
⅞	9	.8028	.7958	.8028	.8028	.8098
1	8	.9188	.9112	.9188	.9188	.9264
1⅛	7	1.0322	1.0237	1.0322	1.0322	1.0407
1¼	7	1.1572	1.1487	1.1572	1.1572	1.1657
1⅜	6	1.2667	1.2566	1.2667	1.2667	1.2768
1½	6	1.3917	1.3816	1.3917	1.3917	1.4018
1¾	5	1.6201	1.6085	1.6201	1.6201	1.6317
2	4½	1.8557	1.8430	1.8557	1.8557	1.8684
2¼	4½	2.1057	2.0930	2.1057	2.1057	2.1184
2½	4	2.3376	2.3236	2.3376	2.3376	2.3516
2¾	4	2.5876	2.5736	2.5876	2.5876	2.6016
3	4	2.8376	2.8236	2.8376	2.8376	2.8516

National Fine Series

Size	Threads per Inch	Basic Pitch Diameter	Screws		Nuts	
			Minimum	Maximum	Minimum	Maximum
0	80	.0519	.0502	.0519	.0519	.0536
1	72	.0640	.0622	.0640	.0640	.0658
2	64	.0759	.0740	.0759	.0759	.0778
3	56	.0874	.0854	.0874	.0874	.0894
4	48	.0985	.0963	.0985	.0985	.1007
5	44	.1102	.1079	.1102	.1102	.1125
6	40	.1218	.1194	.1218	.1218	.1242
8	36	.1460	.1435	.1460	.1460	.1485
10	32	.1697	.1670	.1697	.1697	.1724
12	28	.1928	.1897	.1928	.1928	.1959
¼	28	.2268	.2237	.2268	.2268	.2299
5⁄16	24	.2854	.2821	.2854	.2854	.2887
⅜	24	.3479	.3446	.3479	.3479	.3512
7⁄16	20	.4050	.4014	.4050	.4050	.4086
½	20	.4675	.4639	.4675	.4675	.4711
9⁄16	18	.5264	.5223	.5264	.5264	.5305
⅝	18	.5889	.5848	.5889	.5889	.5930
¾	16	.7094	.7049	.7094	.7094	.7139
⅞	14	.8286	.8237	.8286	.8286	.8335
1	14	.9536	.9487	.9536	.9536	.9585
1⅛	12	1.0709	1.0653	1.0709	1.0709	1.0765
1¼	12	1.1959	1.1903	1.1959	1.1959	1.2015
1⅜	12	1.3209	1.3153	1.3209	1.3209	1.3265
1½	12	1.4459	1.4403	1.4459	1.4459	1.4515

American National Pipe Thread

A = Pitch Diameter of thread at end of pipe
B = Pitch Diameter of thread at gauging notch
D = Outside Diameter of pipe
L' = Normal Engagement by hand between external and internal thread.

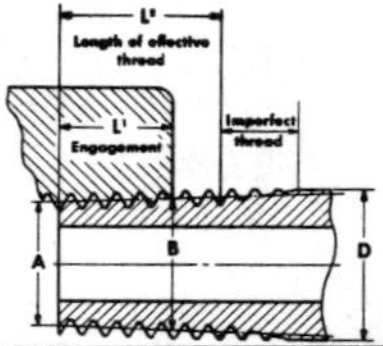

Taper ¾ inch per foot on diameter

Nominal Size Inches	No. of Threads per Inch	Pitch Diameter		Length		Pipe O.D. D Inches	Depth of Thread Inches	Tap Drills for Pipe Threads	
		A Inches	B Inches	L2 Inches	L1 Inches			Minor Diameter Small End of Pipe	Size Drill
⅛	27	.36351	.37476	.2639	.180	.405	.02963	.3339	R
¼	18	.47739	.48989	.4018	.200	.540	.04444	.4329	7⁄16
⅜	18	.61201	.62701	.4078	.240	.675	.04444	.5676	37⁄64
½	14	.75843	.77843	.5337	.320	.840	.05714	.7013	23⁄32
¾	14	.96768	.98887	.5457	.339	1.050	.05714	.9105	59⁄64
1	11½	1.21363	1.23863	.6828	.400	1.315	.06957	1.1441	1 5⁄32
1¼	11½	1.55713	1.58338	.7068	.420	1.660	.06957	1.4876	1½
1½	11½	1.79609	1.82234	.7235	.420	1.900	.06957	1.7265	1 47⁄64
2	11½	2.26902	2.29627	.7565	.436	2.375	.06957	2.1995	2 7⁄32
2½	8	2.71953	2.76216	1.1375	.682	2.875	.10000	2.6195	2⅝
3	8	3.34062	3.38850	1.2000	.766	3.500	.10000	3.2406	3¼
3½	8	3.83750	3.88850	1.2500	.821	4.000	.10000	3.7375	3¾
4	8	4.33438	4.38712	1.3000	.844	4.500	.10000	4.2344	4¼

COUNTERBORE SIZES FOR CAP SCREWS AND MACHINE SCREWS

Size Thread N.F. and N.C.	Fillister Head Cap Screw — Head and Body		Fillister Head Cap Screw — Head and Tap Hole		Body and Tap Hole for any Screw		Round or Hexagon Head Cap Screw — Head and Body		Fillister Head Machine Screw — Head and Body		Round or Hex. Head Machine Screw — Head and Body	
	Cutter for Head	Pilot for Body	Cutter for Head	Pilot for Tap Hole	Cutter for Body	Pilot for Tap Hole	Cutter for Head	Pilot for Body	Cutter for Head	Pilot for Body	Cutter for Head	Pilot for Body
1/4-28	.375	.250	.375	.213	.250	.213	.500	.250	.437	.250	.500	.250
1/4-20	.375	.250	.375	.201	.250	.201	.500	.250	.437	.250	.500	.250
5/16-24	.437	.312	.437	.272	.312	.272	.625	.312	.531	.312	.625	.312
5/16-18	.437	.312	.437	.257	.312	.257	.625	.312	.531	.312	.625	.312
3/8-24	.562	.375	.562	.332	.375	.332	.687	.375	.625	.375	.750	.375
3/8-16	.562	.375	.562	.312	.375	.312	.687	.375	.625	.375	.750	.375
7/16-20	.625	.437	.625	.390	.437	.390	.813	.437	.718	.437	.875	.437
7/16-14	.625	.437	.625	.368	.437	.368	.813	.437	.718	.437	.875	.437
1/2-20	.750	.500	.750	.453	.500	.453	.875	.500	.843	.500	1.000	.500
1/2-13	.750	.500	.750	.421	.500	.421	.875	.500	.843	.500	1.000	.500
9/16-20	.812	.562	.812	.515	.562	.515	1.000	.562				
9/16-12	.812	.562	.812	.484	.562	.484	1.000	.562				
5/8-18	.875	.625	.875	.578	.625	.578	1.062	.625				
5/8-11	.875	.625	.875	.531	.625	.531	1.062	.625				
3/4-16	1.000	.750	1.000	.687	.750	.687	1.312	.750				
3/4-10	1.000	.750	1.000	.656	.750	.656	1.312	.750				
7/8-14	1.125	.875	1.125	.812	.875	.812	1.375	.875				
7/8-9	1.125	.875	1.125	.765	.875	.765	1.375	.875				
1-12	1.312	1.000	1.312	.920	1.000	.920	1.500	1.000				
1-8	1.312	1.000	1.312	.875	1.000	.875	1.500	1.000				

THREAD FORM—ISO METRIC AND ISO INCH SERIES

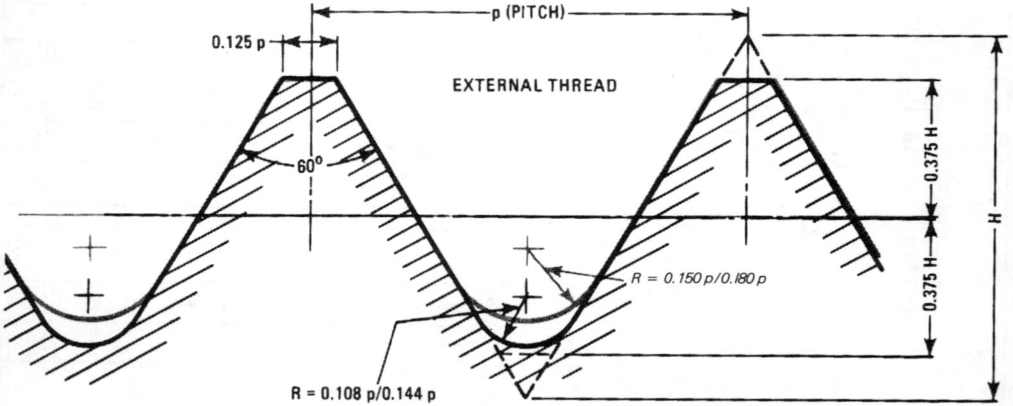

EXTERNAL THREAD

p (PITCH)

H

0.375 H

0.375 H

0.125 p

60°

R = 0.150 p/0.180 p

R = 0.108 p/0.144 p

108

THREAD FORM — RECOMMENDED METRIC SERIES

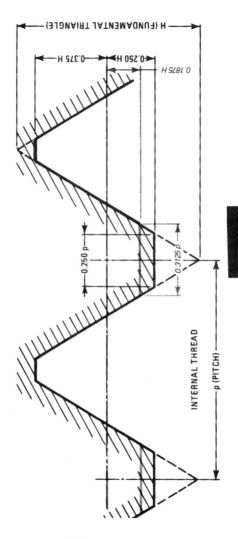

Sec. IV
Screw
Threads

COMPARISON CHART

	INCH SYSTEM				METRIC SYSTEM						
Dia. in Inches	Nominal Size	Dec. Size	Coarse TPI	Fine TPI	Nominal Size	Dec. Eq.(in)	Coarse Pitch MM	Coarse TPI	Fine Pitch MM	Fine TPI	Nominal Size
1.050					M27	1.063	3	8.5	2	12.5	M27x2
1.000	1	1.000	8	12							
.950					M24	.945	3	8.5	2	12.5	M24x2
.900	7/8	.875	9	14							
.850					M22	.866	2.5	10	1.5	17	M22x1.5
.750	3/4	.750	10	16	M20	.787	2.5	10	1.5	17	M20x1.5
.700					M18	.709	2.5	10	1.5	17	M18x1.5
.600	5/8	.625	11	18	M16	.630	2	12.5	1.5	17	M16x1.5
.550					M14	.551	2	12.5	1.5	17	M14x1.5
.500	1/2	.500	13	20							

DIAMETER/THREAD PITCH COMPARISON

COMPARISON CHART

Dia. in Inches	INCH SYSTEM				METRIC SYSTEM						
	Nominal Size	Dec. Size	Coarse TPI	Fine TPI	Nominal Size	Dec. Eq. (in)	Coarse Pitch MM	Coarse TPI	Fine Pitch MM	Fine TPI	Nominal Size
.450	7/16	.437	14	20	M12	.471	1.75	14.5	1.25	20	M12x1.25
.400	3/8	.375	16	24	M10	.393	1.5	17	1.25	20	M10x1.25
.350	5/16	.312	18	24	M8	.315	1.25	20	1.0	25	M8x1
.300	1/4	.250	20	28							
.250					M6	.236	1.0	25	.75	34	M6x.75
	10	.190	24	32	M5	.196	.8	32	.5	51	M5x.5
.200	8	.164	32	36	M4	.157	.7	36	.35	51	M4x.5
	6	.138	32	40	M3	.118	.5	51	.35	74	M3x.35
.150	5	.125	40	44							
	4	.112	40	48	M2.5	.098	.45	56	.35	74	M2.5x.35
	3	.099	48	50	M2	.079	.4	64	.25	101	M2x.25
.100	2	.086	56	64	M1.6	.063	.35	74	2	127	M1.6x.2
	1	.073	64	72	M1.4	.055	.3	85	2	127	M1.4x.2
	0	.060		80							
.050											

DIAMETER/THREAD PITCH COMPARISON

FORMULA FOR OBTAINING TAP DRILL SIZES—METRIC

$$\text{Outside Diam. of Thread} - 1.299 \times \text{mm Pitch} \times \text{Percentage of Full Thread} = \text{Drilled Hole Size}$$

PERCENTAGE OF FULL THREAD FOR OTHER DRILL SIZES

$$\frac{\text{Outside Diam. of thread} - \text{Selected Drill Diam.}}{1.299 \times \text{mm Pitch}} = \text{Amount of percentage of full thread}$$

Figures in table show amount to subtract from O. D. of screw to obtain specific percentage of thread.

EXAMPLE: To find the hole size for obtaining 75% of thread in a 6.5mm – 1.25mm tapped hole, follow first column to 1.25 threads, then across to 75% of thread. This figure (1.2177) when subtracted from the 6.5000 diameter is 5.2823, which is the required diameter of hole.

mm Pitch	Double Depth	50% Thread	55% Thread	60% Thread	65% Thread	70% Thread	75% Thread	80% Thread	85% Thread
4.00	5.1963	2.5982	2.8580	3.1178	3.3776	3.6374	3.8972	4.1570	4.4169
3.50	4.5466	2.2733	2.5006	2.7280	2.9553	3.1826	3.4100	3.6373	3.8646
3.00	3.8969	1.9485	2.1433	2.3381	2.5330	2.7278	2.9227	3.1175	3.3124
2.50	3.2476	1.6238	1.7862	1.9486	2.1109	2.2733	2.4357	2.5981	2.7605

mm Pitch	Double Depth	50% Thread	55% Thread	60% Thread	65% Thread	70% Thread	75% Thread	80% Thread	85% Thread
2.00	2.5979	1.2990	1.4288	1.5587	1.6886	1.8185	1.9484	2.0783	2.2082
1.75	2.2733	1.1367	1.2503	1.3640	1.4776	1.5913	1.7050	1.8186	1.9323
1.50	1.9487	.9744	1.0718	1.1692	1.2667	1.3641	1.4615	1.5590	1.6564
1.25	1.6236	.8118	.8930	.9742	1.0553	1.1365	1.2177	1.2989	1.3801
1.00	1.2990	.6495	.7145	.7794	.8444	.9093	.9743	1.0392	1.1042
.90	1.1687	.5844	.6428	.7012	.7597	.8181	.8765	.9350	.9934
.80	1.0394	.5197	.5717	.6236	.6756	.7276	.7796	.8315	.8835
.75	.9743	.4871	.5359	.5846	.6333	.6820	.7307	.7794	.8282
.70	.9093	.4547	.5001	.5456	.5910	.6365	.6820	.7274	.7729
.60	.7793	.3897	.4286	.4676	.5065	.5455	.5845	.6234	.6624
.50	.6421	.3211	.3532	.3853	.4174	.4495	.4816	.5137	.5458
.45	.5847	.2924	.3216	.3508	.3801	.4093	.4385	.4678	.4970
.40	.5197	.2599	.2858	.3118	.3378	.3638	.3898	.4158	.4417
.35	.4547	.2274	.2501	.2728	.2956	.3183	.3410	.3638	.3865
.30	.3896	.1948	.2143	.2338	.2532	.2727	.2922	.3117	.3312
.25	.3246	.1663	.1785	.1948	.2110	.2272	.2434	.2597	.2759

MEASURING METRIC THREADS
by
THREE WIRE METHOD
(use chart on opposite page)

Notation
M = Measurement over wires—(M = P.D. + C)
D = Outside diameter of thread in mm
P.D. = Pitch diameter in mm—(P.D. = M − C)
p = Pitch in mm
W = Diameter of best wire size in mm from chart
C = Constant from chart

Note
Metric threads are designated by the basic outside diameter and the pitch of the thread in mm. Thus, a M12 X 1.5 metric thread means that D = 12mm and p = 1.5 mm.

Example
To find the measurement over the wires (M) for a M12 X 1.5 metric thread, proceed as follows:

First: Select the proper wire size from the chart for the particular pitch of thread.
          W = Wire size for 1.5 mm pitch = .866 mm.

Second: Find the basic depth of thread.
          d = p X .64952 = 1.5 X .64952 = .97428.

Third: Find the basic pitch diameter.
          P.D. = D − d = 12 − .97428 = 11.02572.

Fourth: Select constant from chart.
          C = 1.299.

Fifth: Find the measurement over the wires.
          M = P.D. + C = 11.02572 + 1.299
          = 12.32472.

METRIC THREAD MEASURING WIRES

Pitch mm	Best Wire		Constant	
	mm	Inch equivalent	mm	Inch equivalent
0.2	0.1155	.00455	0.1733	.00683
0.225	0.1299	.00511	0.1948	.00766
0.25	0.1443	.00568	0.2164	.00852
0.3	0.1732	.00682	0.2598	.01023
0.35	0.2021	.00796	0.3032	.01195
0.4	0.2309	.00909	0.3463	.01363
0.45	0.2598	.01023	0.3897	.01535
0.5	0.2887	.01137	0.4331	.01706
0.6	0.3464	.01364	0.5196	.02046
0.7	0.4041	.01591	0.6061	02386
0.75	0.433	.01705	0.6495	.02558
0.8	0.4619	.01818	0.6929	.02726
0.9	0.5197	.02046	0.7794	.03069
1	0.5774	.02273	0.8662	.03409
1.25	0.7217	.02841	1.0826	.04261
1.5	0.866	.03410	1.299	.05116
1.75	1.0104	.03978	1.5157	.05967
2	1.1547	.04546	1.732	.06819
2.5	1.4434	.05683	2.1651	.08525
3	1.7321	.06819	2.5982	.10228
3.5	2.0207	.07956	3.031	.11935
4	2.3094	.09092	3.4641	.13638
4.5	2.5981	.10229	3.8972	.15344
5	2.8868	.11365	4.3303	.17047
5.5	3.1754	.12502	4.7631	.18753
6	3.4641	.13638	5.1961	.20457
7	4.0415	.15911	6.0623	.23866
8	4.6188	.18184	6.9282	.27276
9	5.1962	.20457	7.7944	.30685
10	5.7735	.22730	8.6602	.34095

Sec. IV
Screw
Threads

THE THREE-WIRE METHOD

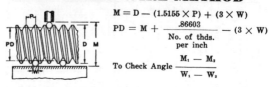

$$M = D - (1.5155 \times P) + (3 \times W)$$

$$PD = M + \frac{.86603}{\text{No. of thds. per inch}} - (3 \times W)$$

To Check Angle $\dfrac{M_1 - M_2}{W_1 - W_2}$

M = Measurement over best size wire.
M_1 = Measurement over maximum size wire.
M_2 = Measurement over minimum size wire.
D = Outside Diameter of Thread.
P.D. = Pitch Diameter.
W = Diameter Best size wire.   0.57735 × pitch
W_1 = Diameter maximum size wire.   0.90 × pitch.
W_2 = Diameter minimum size wire.   0.56 × pitch

Three Wire Measurement of American (National) Std.

No. Thds. per inch	Pitch Thds. per inch	Best Wire Size .57735 x Pitch	Maximum Wire Size	Minimum Wire Size
4	.250000	.144337	.225000	.140000
4½	.222222	.128300	.200000	.124444
5	.200000	.115470	.180000	.112000
5½	.181818	.104969	.163636	.101818
6	.166666	.096224	.149999	.093333
7	.142857	.082478	.128571	.080000
7½	.133333	.076979	.120000	.074666
8	.125000	.072168	.112500	.070000
9	.111111	.064149	.100000	.062222
10	.100000	.057735	.090000	.056000
11	.090999	.052486	.081818	.050909
11½	.086956	.050204	.078260	.048695
12	.083333	.048112	.075000	.046666
13	.076923	.044411	.069231	.043077
14	.071428	.041239	.064285	.040000
16	.062500	.036084	.056250	.035000
18	.055555	.032074	.050000	.031111
20	.050000	.028867	.045000	.028000
22	.045454	.026242	.040909	.025454
24	.041666	.024055	.037499	.023333
26	.038461	.022205	.034615	.021538
27	.037037	.021383	.033333	.022543
28	.035714	.020620	.032143	.020000
30	.033333	.019244	.030000	.018666
32	.031250	.018042	.028125	.017500
36	.027777	.016037	.024999	.015555
40	.025000	.014433	.022500	.014000
44	.022727	.013121	.020454	.014727
48	.020833	.012027	.018750	.011666
50	.020000	.011547	.018000	.011200
56	.017857	.010309	.016071	.010000
64	.015625	.009021	.014063	.008750
72	.0138888	.008018	.012499	.007777
80	.012500	.007216	.011250	.007000

METRIC PITCH CONVERSIONS
MILLIMETER PITCH TO THREADS PER INCH

Pitch In Millimeters	Pitch In Inches	Threads Per In.	Basic Height	Pitch In Millimeters	Pitch In Inches	Threads Per In.	Basic Height
.25	.00984	101.6000	.00639	1.25	.04921	20.3211	.03196
.30	.01181	84.6668	.00767	1.50	.05906	16.9316	.03836
.35	.01378	72.5689	.00895	1.75	.06890	14.5138	.04475
.40	.01575	63.4921	.01023	2.00	.07874	12.7000	.05114
.45	.01772	56.4334	.01151	2.50	.09843	10.1595	.06393
.50	.01969	50.8000	.01279	3.00	.11811	8.4667	.07671
.60	.02362	42.3370	.01534	3.50	.13780	7.2569	.08950
.70	.02756	36.2845	.01790	4.00	.15748	6.3500	.10229
.75	.02953	33.8639	.01918	4.50	.17717	5.6443	.11508
.80	.03150	31.7460	.02046	5.00	.19685	5.0800	.12785
.90	.03543	28.2247	.02301	6.00	.23622	4.2333	.15344
1.00	.03937	25.4000	.02557				

117

NOTES

Feeds and Speeds For Milling

(Average Conditions with High Speed Steel Cutters)

DEPTH OF CUT IN INCHES		WIDTH OF CUT IN INCHES											
		1/4	1/2	3/4	1	1 1/4	1 1/2	2	2 1/2	3	4	5	6
1/4	Feed	4.5	4.5	4	4	4	3.5	3.5	3.5	3	3	3	2.5
	Speed	110	110	105	105	100	100	100	95	95	95	90	90
1/2	Feed	4.5	4	4	4	3.5	3.5	3.5	3	3	3	2.5	2.5
	Speed	105	105	100	100	95	95	95	90	90	90	85	85
3/4	Feed	4	4	3.5	3.5	3.5	3	3	3	2.5	2.5	2.25	2.25
	Speed	100	100	95	95	90	90	90	85	85	85	80	80
1	Feed	4	3.5	3.5	3	3	3	3	2.5	2.5	2.5	2.25	2
	Speed	95	95	90	90	85	85	85	80	80	80	75	75
1 1/4	Feed	3.5	3.5	3.5	3	3	3	2.5	2.5	2.25	2	1.75	1.5
	Speed	90	90	85	85	80	80	80	75	75	75	70	70
1 1/2	Feed	3.5	3.5	3	3	2.5	2.5	2.5	2.25	2	1.75	1.5	1.25
	Speed	85	85	80	80	75	75	75	70	70	70	65	65
2	Feed	3.5	3	3	3	2.5	2.5	2.25	2	1.75	1.75	1.25	1
	Speed	80	80	75	75	70	70	70	65	65	65	60	60

Correction Factors

S.A.E. 1020 — Annld. = 1.00
S.A.E. 1035 — Annld. = .80
Tool Steel — Annld. = .60
Alloy Steels:
  Free Machining = .80
  Medium = .65
  Tough = .50
Cast Steel = .40
Cast Irons:
  Hard = .70
  Medium = .95
  Soft = 1.20
Malleable Iron = .80
Brass and Bronze:
  Free Machining = 2.50
  Medium = 1.60
  Hard = .80
Monel Metal = .70
Magnesium Alloys = 5.00
Aluminum = 2.00

Feeds are given in Inches per Minute.
Speeds are given in Surface Feet per Minute.
To obtain the proper Feed and Speed for a given material, multiply both Feed and Speed figures given in table by the correction factor for that material.

Sec. V
Milling,
Shaping,
Turning

SPEED AND FEED CALCULATIONS
For Milling Cutters and Other Rotating Tools

TO FIND	HAVING	FORMULA
Surface (or Periphery) Speed in Feet per Minute = S.F.M.	D = Diameter of Tool in Inches and R.P.M. = Revolutions per Minute	$S.F.M. = \dfrac{D \times 3.1416 \times R.P.M.}{12}$
Revolutions per Minute = R.P.M.	S.F.M. = Surface Speed in Feet per Minute and D = Diameter of Tool in Inches	$R.P.M. = \dfrac{S.F.M. \times 12}{D \times 3.1416}$
Feed per Revolution in Inches = F.R.	F.M. = Feed in Inches per Minute and R.P.M. = Revolutions per Minute	$F.R. = \dfrac{F.M.}{R.P.M.}$
Feed in Inches per Minute = F.M.	F.R. = Feed per Revolution in Inches and R.P.M. = Revolutions per Minute	$F.M. = F.R. \times R.P.M.$
Number of Cutting Teeth per Minute = T.M.	T = Number of Teeth in Tool and R.P.M. = Revolutions per Minute	$T.M. = T \times R.P.M.$
Feed per Tooth = F.T.	T = Number of Teeth in Tool and F.R. = Feed per Revolution in Inches	$F.T. = \dfrac{F.R.}{T}$
Feed per Tooth = F.T.	T = Number of Teeth in Tool F.M. = Feed in Inches per Minute and R.P.M. = Speed in Revolutions per Minute	$F.T. = \dfrac{F.M.}{T \times R.P.M.}$

SPEED AND FEED CALCULATIONS—METRIC
For Milling Cutters and Other Rotating Tools

To Find	Having	Formulas
Surface (periphery) speed in meters per minute S = m/min	D = Diameter of tool in millimeters R.P.M. = Revolutions per minute	$S = m/min = D \times .00314 \times R.P.M.$
Revolutions per minute = R.P.M.	S = Surface speed in meters per minute D = Diameter of tool in millimeters	$R.P.M. = \dfrac{m/min}{D \times .00314}$
Feed per revolution in millimeters = F.R.	F.M. = Feed in millimeters per minute R.P.M. = Revolutions per minute	$F.R. = \dfrac{F.M.}{R.P.M.}$
Feed in millimeters per minute = F.M.	F.R. = Feed per revolution in millimeters R.P.M. = Revolutions per minute	$F.M. = F.R. \times R.P.M.$
Number of cutting teeth per minute = T.M.	T. = Number of teeth in tool R.P.M. = Revolutions per minute	$T.M. = T \times R.P.M.$
Feed per tooth = F.T.	T. = Number of teeth in tool R.P.M. = Revolutions per minute	$F.T. = \dfrac{F.R.}{T.}$
Feed per tooth = F.T.	T. = Number of teeth in tool F.M. = Feed in millimeters per minute R.P.M. = Revolutions per minute	$F.T. = \dfrac{F.M.}{T. \times R.P.M.}$

Sec. V
Milling,
Shaping,
Turning

Sharpening Milling Cutters

Setting the Tooth Rest, Using a Straight Wheel

Figure A illustrates a milling cutter being ground with a straight wheel. The distance C between the center lines of the wheel and cutter varies with the clearance angle. The method of producing the desired clearance angle, when using a straight wheel, is as follows:

1. Bring the center of the wheel and the work into the same plane.
2. Fasten the tooth rest to the table of the machine and adjust the tooth rest to the same height as the center of the work, using a height gauge.
3. Raise (or lower, depending upon the direction of wheel rotation) the wheel head the proper distance by means of the graduated hand wheel (on Norton tool and cutter grinding machines). On some machines, where the wheel head is stationary and the table moves, the same effect is obtained by lowering the table.

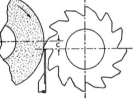

Figure A — Developing the clearance angle when using a straight wheel

The distance to raise or lower the wheel head when using a straight wheel may be calculated as follows: Multiply the clearance angle in degrees by the diameter of the wheel in inches, and this product by the constant .0087.

> *Example:* To determine the distance to raise or lower the wheel head for a 7° clearance angle, using a straight wheel 6″ diameter.
> *Solution:* C = 7° x 6″ x .0087 = .365″.

Figure B shows the use of a cup wheel for grinding a cutter. The setting for producing the desired clearance angle, of a spiral milling cutter, for example, is obtained as follows:

1. Fasten the tooth rest to the wheel head and line up the tooth rest and center of the cutter in the same plane, using a height gauge.

2. Raise or lower the wheel head with the tooth rest the required distance.

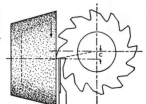

Figure B –Developing the clearance angle when using a cup wheel

To calculate this distance when using a cup wheel, multiply the required clearance angle by the diameter of the cutter in inches and this product by the constant .0087.

Example: To determine the distance to raise or lower the wheel head beyond the center of a cutter 3″ in diameter in order to produce a clearance angle of 5°, using a cup wheel 4″ in diameter.
Solution: C = 5° x 3″ x .0087 = .130″.

As a general rule, the clearance angle should be from 6° to 7° for cutters up to 3″ in diameter and from 4° to 5° for cutters larger than 3″. The following table may also be used as a guide in selecting the proper clearance angle according to the material to be cut:

Low Carbon Steels	5° to 7°
High Carbon and Alloy Steels	3° to 5°
Steel Castings	5° to 7°
Cast Iron	4° to 7°
Brass and Soft Bronze	10° to 12°
Medium and Hard Bronze	4° to 7°
Aluminum	10° to 12°

Clearance Table—Cup Wheels

C = Distance in inches to set tip of tooth rest below or above center of cutter when grinding the peripheral teeth of cutters with a cup wheel.

Cutter Diameter (Inches)	C for 4° Clearance	C for 5° Clearance	C for 6° Clearance	C for 7° Clearance
½	.017	.022	.026	.031
¾	.026	.033	.040	.046
1	.035	.044	.053	.061
1¼	.044	.055	.066	.077
1½	.053	.066	.079	.092
1¾	.061	.076	.092	.108
2	.070	.087	.105	.123
2¼	.087	.109	.131	.153
2¾	.097	.120	.144	.168
3	.105	.131	.158	.184
3½	.122	.153	.184	.215
4	.140	.195	.210	.245
4½	.157	.197	.237	.276
5	.175	.219	.263	.307
5½	.192	.241	.289	.338
6	.210	.262	.315	.368

Clearance Table—Straight Wheels

C = Distance in inches to set center of cutter and tip of tooth rest below or above center of wheel when grinding with a straight wheel.

Wheel Diameter (Inches)	C for 4° Clearance	C for 5° Clearance	C for 6° Clearance	C for 7° Clearance
3	.105	.131	.158	.184
3¼	.113	.142	.170	.199
3½	.122	.153	.184	.215
3¾	.131	.164	.197	.230
4	.140	.175	.210	.245
4¼	.150	.185	.223	.263
4½	.157	.197	.236	.276
4¾	.166	.207	.249	.291
5	.175	.219	.263	.307
5¼	.183	.229	.275	.322
5½	.192	.241	.289	.338
5¾	.201	.251	.302	.352
6	.210	.262	.315	.368
6¼	.218	.273	.328	.383
6½	.227	.284	.342	.399
6¾	.236	.295	.355	.414
7	.245	.306	.368	.430

Sec. V
Milling,
Shaping,
Turning

INDEX TABLE 2 to 50

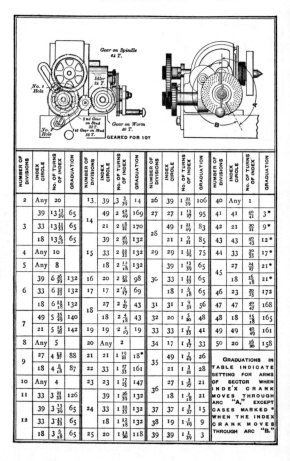

GEARED FOR 107

No. of Divisions	Index Circle	No. of Turns of Index	Graduation
2	Any	20	
3	39	$13\frac{13}{39}$	65
3	33	$13\frac{11}{33}$	65
3	18	$13\frac{6}{18}$	65
4	Any	10	
5	Any	8	
6	39	$6\frac{36}{39}$	132
6	33	$6\frac{22}{33}$	132
6	18	$6\frac{12}{18}$	132
7	49	$5\frac{35}{49}$	140
7	21	$5\frac{15}{21}$	142
8	Any	5	
9	27	$4\frac{12}{27}$	88
9	18	$4\frac{8}{18}$	87
10	Any	4	
11	33	$3\frac{21}{33}$	126
12	39	$3\frac{13}{39}$	65
12	33	$3\frac{11}{33}$	65
12	18	$3\frac{6}{18}$	65
13	39	$3\frac{3}{39}$	14
14	49	$2\frac{42}{49}$	169
14	21	$2\frac{18}{21}$	170
14	39	$2\frac{26}{39}$	132
15	33	$2\frac{22}{33}$	132
15	18	$2\frac{12}{18}$	132
16	20	$2\frac{10}{20}$	98
17	17	$2\frac{6}{17}$	69
18	27	$2\frac{6}{27}$	43
18	18	$2\frac{4}{18}$	43
19	19	$2\frac{2}{19}$	19
20	Any	2	
21	21	$1\frac{19}{21}$	18*
22	33	$1\frac{27}{33}$	161
23	23	$1\frac{17}{23}$	147
24	39	$1\frac{26}{39}$	132
24	33	$1\frac{22}{33}$	132
24	18	$1\frac{12}{18}$	132
25	20	$1\frac{15}{20}$	118
26	39	$1\frac{21}{39}$	106
27	27	$1\frac{13}{27}$	95
28	49	$1\frac{21}{49}$	83
28	21	$1\frac{9}{21}$	85
29	29	$1\frac{15}{29}$	75
30	39	$1\frac{13}{39}$	65
30	33	$1\frac{11}{33}$	65
30	18	$1\frac{6}{18}$	65
31	31	$1\frac{9}{31}$	56
32	20	$1\frac{5}{20}$	48
33	33	$1\frac{7}{33}$	41
34	17	$1\frac{3}{17}$	33
35	49	$1\frac{7}{49}$	26
35	21	$1\frac{3}{21}$	28
36	27	$1\frac{3}{27}$	21
36	18	$1\frac{2}{18}$	21
37	37	$1\frac{3}{37}$	15
38	19	$1\frac{1}{19}$	9
39	39	$1\frac{1}{39}$	3
40	Any	1	
41	41	$\frac{40}{41}$	3*
42	21	$\frac{20}{21}$	9*
43	43	$\frac{40}{43}$	12*
44	33	$\frac{30}{33}$	17*
45	27	$\frac{24}{27}$	21*
45	18	$\frac{16}{18}$	21*
46	23	$\frac{20}{23}$	172
47	47	$\frac{40}{47}$	168
48	18	$\frac{15}{18}$	165
49	49	$\frac{40}{49}$	161
50	20	$\frac{16}{20}$	158

GRADUATIONS IN TABLE INDICATE SETTING FOR ARMS OF SECTOR WHEN INDEX CRANK MOVES THROUGH ARC "A," EXCEPT CASES MARKED * WHEN THE INDEX CRANK MOVES THROUGH ARC "B."

Sec. V
Milling,
Shaping,
Turning

INDEX TABLE 51 to 92.

NUMBER OF DIVISIONS	INDEX CIRCLE	NO. OF TURNS OF INDEX	GRADUATION	GEAR ON WORM	1ST GEAR ON STUD (NO.1 HOLE)	2ND GEAR ON STUD (NO.1 HOLE)	GEAR ON SPINDLE	IDLERS NO.1 HOLE	IDLERS NO.2 HOLE
51	17	$\frac{14}{17}$	33*	24			48	24	44
52	39	$\frac{30}{39}$	152						
53	49	$\frac{35}{49}$	140	56	40	24	72		
53	21	$\frac{15}{21}$	142	56	40	24	72		
54	27	$\frac{20}{27}$	147						
55	33	$\frac{24}{33}$	144						
56	49	$\frac{35}{49}$	140						
56	21	$\frac{15}{21}$	142						
57	49	$\frac{35}{49}$	140	56			40	24	44
57	21	$\frac{15}{21}$	142	56			40	24	44
58	29	$\frac{20}{29}$	136						
59	39	$\frac{26}{39}$	132	48			32		44
59	33	$\frac{22}{33}$	132	48			32		44
59	18	$\frac{12}{18}$	132	48			32		44
60	39	$\frac{26}{39}$	132						
60	33	$\frac{22}{33}$	132						
60	18	$\frac{12}{18}$	132						
61	39	$\frac{26}{39}$	132	48			32	24	44
61	33	$\frac{22}{33}$	132	48			32	24	44
61	18	$\frac{12}{18}$	132	48			32	24	44
62	31	$\frac{20}{31}$	127						
63	39	$\frac{26}{39}$	132	24			48	24	44
63	33	$\frac{22}{33}$	132	24			48	24	44
63	18	$\frac{12}{18}$	132	24			48	24	44
64	16	$\frac{10}{16}$	123						
65	39	$\frac{24}{39}$	121						
66	33	$\frac{20}{33}$	120						
67	49	$\frac{35}{49}$	112	28			48		44
67	21	$\frac{12}{21}$	113	28			48		44
68	17	$\frac{10}{17}$	116						
69	20	$\frac{12}{20}$	118	40			56	24	44
70	49	$\frac{28}{49}$	112						
70	21	$\frac{12}{21}$	113						
71	27	$\frac{15}{27}$	110	72			40	24	
71	18	$\frac{10}{18}$	109	72			40	24	
72	27	$\frac{15}{27}$	110						
72	18	$\frac{10}{18}$	109						
73	49	$\frac{28}{49}$	112	28			48	24	44
73	21	$\frac{12}{21}$	113	28			48	24	44
74	37	$\frac{20}{37}$	107						
75	15	$\frac{8}{15}$	105						
76	19	$\frac{10}{19}$	103						
77	20	$\frac{10}{20}$	98	32			48		44
78	39	$\frac{20}{39}$	101						
79	20	$\frac{10}{20}$	98	48			24		44
80	20	$\frac{10}{20}$	98						
81	20	$\frac{10}{20}$	98	48			24	24	44
82	41	$\frac{20}{41}$	96						
83	26	$\frac{10}{20}$	98	32			48	24	44
84	21	$\frac{10}{21}$	94						
85	17	$\frac{8}{17}$	92						
86	43	$\frac{20}{43}$	91						
87	15	$\frac{7}{15}$	92	40			24	24	44
88	33	$\frac{15}{33}$	89						
89	27	$\frac{12}{27}$	88	72			32		44
89	18	$\frac{8}{18}$	87	72			32		44
90	27	$\frac{12}{27}$	88						
90	18	$\frac{8}{18}$	87						
91	39	$\frac{18}{39}$	91	24			48	24	44
92	23	$\frac{10}{23}$	86						

SHAPER TOOL SHAPES

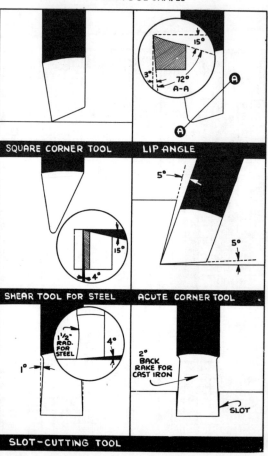

SQUARE CORNER TOOL

LIP ANGLE

15°
3°
72°
A-A

A

A

SHEAR TOOL FOR STEEL

15°
4°

ACUTE CORNER TOOL

5°
5°

SLOT-CUTTING TOOL

1½" RAD. FOR STEEL
4°
1°

2° BACK RAKE FOR CAST IRON
SLOT

SHAPER TOOL SHAPES

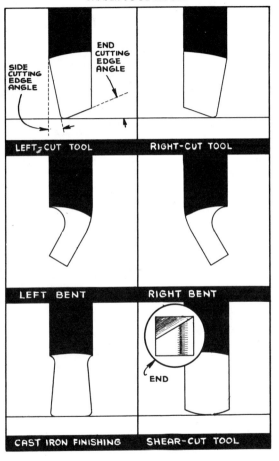

SIDE CUTTING EDGE ANGLE

END CUTTING EDGE ANGLE

LEFT-CUT TOOL

RIGHT-CUT TOOL

LEFT BENT

RIGHT BENT

END

CAST IRON FINISHING

SHEAR-CUT TOOL

Sec. V
Milling,
Shaping,
Turning

129

SHAPER CUTTING SPEEDS – FEET PER MINUTE

(high-speed steel tools)

Material	Roughing	Finishing	Material	Roughing	Finishing
Cast iron	60	100	Hard bro.	60	100
.10 to .20 C	80	120	Brass	150	Max. speed
.20 to .40 C	60	100	Aluminum	150	Max. speed
Die steel	40	40			

CUTTING SPEEDS IN FEET PER MINUTE

16" HIGH SPEED

LENGTH OF STROKE IN INCHES

STROKES PER MIN.	1	2	3	4	5	6	7	8	9	10	11	12	13	14	15	16
15	4	6	7	10	13	16	19	20	23	25	28	29	30	33	35	36
22	6	9	12	15	17	20	23	26	29	32	35	38	41	43	46	49
33	7	12	17	22	28	32	38	42	46	51	55	59	64	67	71	75
45	9	16	22	29	36	42	49	55	61	67	74	80	84	90	96	101
68	12	23	35	44	54	64	74	84	94	103	112	120	129	136	145	154
99	16	33	48	64	78	93	107	120	135	146	159	172	185	197	208	220
145	25	48	71	93	114	136	158	178	198	218	236					
200	34	65	97	127	158	187	216	245								

Note: To use the above chart, read down from the length of stroke in inches and across from the recommended feet per minute to get the number of strokes per minute.

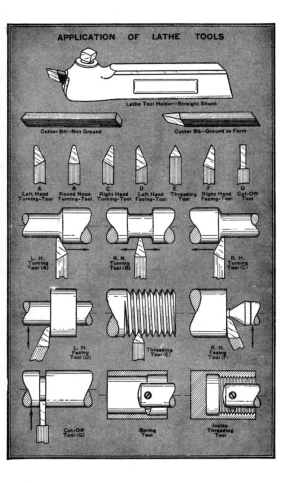

APPLICATION OF LATHE TOOLS

Lathe Tool Holder—Straight Shank

Cutter Bit—Not Ground

Cutter Bit—Ground to Form

A
Left Hand
Turning-Tool

B
Round Nose
Turning-Tool

C
Right Hand
Turning-Tool

D
Left Hand
Facing-Tool

E
Threading
Tool

F
Right Hand
Facing-Tool

G
Cut-Off
Tool

L. H.
Turning
Tool (A)

R. N.
Turning
Tool (B)

R. H.
Turning
Tool (C)

L. H.
Facing
Tool (D)

Threading
Tool (E)

R. H.
Facing
Tool (F)

Cut-Off
Tool (G)

Boring
Tool

Inside
Threading
Tool

SPINDLE SPEEDS IN R.P.M. FOR TURNING AND BORING
Calculated for Average Cuts with High Speed Steel Cutter Bits

Diameter in Inches	Alloy Steels 50 f. p. m.	Cast Iron 75 f. p. m.	Machine Steel 100 f. p. m.	Hard Brass 150 f. p. m.	Soft Brass 200 f. p. m.	Aluminum 300 f. p. m.
1	191	287	382	573	764	1146
2	95	143	191	287	382	573
3	64	95	127	191	254	381
4	48	72	95	143	190	285
5	38	57	76	115	152	228
6	32	48	64	95	128	192
7	27	41	55	82	110	165
8	24	36	48	72	96	144
9	21	32	42	64	84	126
10	19	29	38	57	76	114
11	17	26	35	52	70	105
12	16	24	32	48	64	96
13	15	22	29	44	58	87
14	14	20	27	41	54	81
15	13	19	25	38	50	75
16	12	18	24	36	48	72

The most efficient cutting speed for turning varies with the kind of metal being machined, the depth of the cut, the feed and the type of cutter bit used. If too slow a cutting speed is used, much time may be lost, and if too high a speed is used the tool will dull quickly. The following cutting speeds are recommended for high speed steel cutter bits:

CUTTING SPEEDS IN SURFACE FEET PER MINUTE

Kind of Metal	Roughing Cuts .010 in. to .020 in. Feed	Finishing Cuts .002 in. to .010 in. Feed	Cutting Screw Threads
Cast Iron	60 f. p. m.	80 f. p. m.	25 f. p. m.
Machine Steel	90 f. p. m.	100 f. p. m.	35 f. p. m.
Tool Steel, Annealed ...	50 f. p. m.	75 f. p. m.	20 f. p. m.
Brass	150 f. p. m.	200 f. p. m.	50 f. p. m.
Aluminum	200 f. p. m.	300 f. p. m.	50 f. p. m.
Bronze	90 f. p. m.	100 f. p. m.	25 f. p. m.

If a cutting lubricant is used, the above speeds may be increased 25% to 50%. When using tungsten-carbide tipped cutting tools, the cutting speeds may be increased from 100% to 800%.

CUTTING SPEEDS AND TOOL ANGLES FOR NON-FERROUS MATERIALS

Material	Cutting Speed f. p. m.	Front Clearance Degrees	Side Clearance Degrees	Back Rake Degrees	Side Rake Degrees
Aluminum*.....................	300–400	7	8	30	20
Brass — Leaded................	300–700	6	7	0	0
Bronze — Free Cutting.........	300–700	6	5	0	1½
Bronze — Readily Machinable...	150–300	10	8	7	7
Bronze — Tough...............	75–150	12	15	15	25
Copper — Pure.................	75–150	7	7	10	25
Die Castings..................	225–350	7	7	8	10
Magnesium Alloy	275–400	8	8	6	4
Monel Metal..................	50–170	6	6	7	12
Plastics — Cast Resin...........	200–600	10	12	30	25
Plastics — Cold Set Molded.....	200–600	10	10	0	0
Plastics — Hot Set Molded......	200–600	6	6	10	30
Plastics — Laminated..........	200–600	8	8	30	30
Stainless Steel................	50–150	8	8	9	0
Wood........................	400–800	20	20	30	30

*Kerosene lubricant used.

HOW TO USE NOMOGRAPH METHOD FOR CALCULATING SPEEDS REQUIRED TO MACHINE STEEL WITH CARBIDES

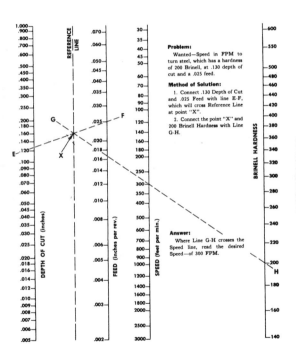

Problem:

Wanted—Speed in FPM to turn steel, which has a hardness of 200 Brinell, at .130 depth of cut and a .025 feed.

Method of Solution:

1. Connect .130 Depth of Cut and .025 Feed with line E-F, which will cross Reference Line at point "X".

2. Connect the point "X" and 200 Brinell Hardness with Line G-H.

Answer:

Where Line G-H crosses the Speed line, read the desired Speed—of 300 FPM.

Sec. V
Milling,
Shaping,
Turning

133

Lubricants for Cutting Tools

Cutting Lubricants are used to keep the tools cool and to reduce the wear on the cutting edges, permitting fast cutting speeds.

Soluble Oil is the medium most generally used and is successful in cutting most grades of steel. It is a saponified oil, usually of a mineral oil base and readily mixes with water, the proportions of oil and water varying with the requirements.

Lard oil is highly efficient in specific cases when used in its pure or un-adulterated state. It is however sometimes mixed with other oils or chemicals, for economic reasons, when it is not necessary to use it pure. Often in combination with sulphur it is highly efficient in tapping hard or tough metals. Such lubricant is called "Sulphur Base Oil."

The chart below gives an idea of the uses of cutting lubricant or coolant.

RECOMMENDED OILS AND CUTTING FLUIDS FOR VARIOUS MATERIALS

Material	Drilling	Reaming	Turning	Milling	Threading
Aluminum	Kerosene Kerosene and Lard Oil Soluble Oil	Kerosene Soluble Oil Mineral Oil	Soluble Oil	Soluble Oil Lard Oil Mineral Oil Dry	Soluble Oil Kerosene and Lard Oil
Brass	Dry Soluble Oil Kerosene and Lard Oil	Dry Soluble Oil	Soluble Oil	Dry Soluble Oil	Lard Oil Soluble Oil
Bronze	Soluble Oil Lard Oil Mineral Oil Dry	Soluble Oil Lard Oil Mineral Oil Dry	Soluble Oil	Soluble Oil Lard Oil Mineral Oil Dry	Lard Oil Soluble Oil
Cast Iron	Dry Air Jet Soluble Oil	Soluble Oil Mineral Lard Oil	Dry Soluble Oil	Dry Soluble Oil	Sulphurized Oil Mineral Lard Oil
Cast Steel	Soluble Oil Mineral Lard Oil Sulphurized Oil	Soluble Oil Mineral Lard Oil Lard Oil	Soluble Oil	Soluble Oil Mineral Lard Oil	Mineral Lard Oil
Copper	Soluble Oil Dry Mineral Lard Oil Kerosene	Soluble Oil Lard Oil	Soluble Oil	Soluble Oil Dry	Soluble Oil Lard Oil
Malleable Iron	Dry Soda Water	Dry Soda Water	Soluble Oil	Dry Soda Water	Lard Oil and Soda
Monel Metal	Lard Oil Soluble Oil	Lard Oil Soluble Oil	Soluble Oil	Soluble Oil	Lard Oil
Steel, Mild	Soluble Oil Mineral Lard Oil Sulphurized Oil Lard Oil	Soluble Oil Mineral Lard Oil	Soluble Oil	Soluble Oil Mineral Lard Oil	Soluble Oil Mineral Lard Oil
Steel, Alloys, Forgings	Soluble Oil Sulphurized Oil Mineral Lard Oil	Soluble Oil Mineral Lard Oil Sulphurized Oil	Soluble Oil	Soluble Oil Mineral Lard Oil	Sulphurized Oil Lard Oil
Steel, Tool	Soluble Oil Mineral Lard Oil Sulphurized Oil	Soluble Oil Lard Oil Sulphurized Oil	Soluble Oil	Soluble Oil Lard Oil	Sulphurized Oil Lard Oil
Steel, Manganese (12 to 15%)	Dry				
Wrought Iron	Soluble Oil Mineral Lard Oil Sulphurized Oil	Soluble Oil Mineral Lard Oil	Soluble Oil	Soluble Oil Mineral Lard Oil	Soluble Oil Mineral Lard Oil

TABLE SHOWING DEPTH OF SPACE AND THICKNESS OF TOOTH IN SPUR GEARS WHEN CUT WITH PATENT CUTTERS

Diametral Pitch of Cutter	Depth to be Cut in Gear	Thickness of Tooth at Pitch Line	Diametral Pitch of Cutter	Depth to be Cut in Gear	Thickness of Tooth at Pitch Line
1¼	1.7257″	1.2566″	11	.1961″	.1428″
1½	1.4381	1.0472	12	.1798	.1309
1¾	1.2326	.8976	14	.1541	.1122
2	1.0785	.7854	16	.1348	.0982
2¼	.9587	.6981	18	.1198	.0873
2½	.8628	.6283	20	.1079	.0785
2¾	.7844	.5712	22	.0980	.0714
3	.7190	.5236	24	.0898	.0654
3½	.6163	.4488	26	.0829	.0604
4	.5393	.3927	28	.0770	.0561
5	.4314	.3142	30	.0719	.0524
6	.3595	.2618	32	.0674	.0491
7	.3081	.2244	36	.0599	.0436
8	.2696	.1963	40	.0539	.0393
9	.2397	.1745	48	.0449	.0327
10	.2157	.1571			

Involute Gear Cutters for Teeth of Gears

Eight Cutters are made for each pitch, as follows:

No. 1	will Cut Gears from	135 teeth	to a rack	
2	" " "	55 " "	134 Teeth
3	" " "	35 " "	54 "
4	" " "	26 " "	34 "
5	" " "	21 " "	25 "
6	" " "	17 " "	20 "
7	" " "	14 " "	16 "
8	" " "	12 " "	13 "

RULES AND FORMULAE FOR SPUR GEAR CALCULATIONS

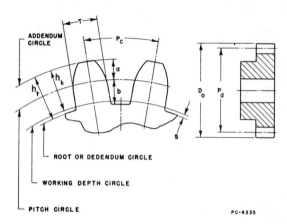

PC-4335

The following symbols are used in conjunction with the formulae for determining the proportions of spur gear teeth.

P = Diametral pitch.
P_c = Circular pitch.
P_d = Pitch diameter.
D_o = Outside diameter.
N = Number of teeth in the gear.
T = Tooth thickness.
a = Addendum.
b = Dedendum.
h_k = Working depth.
h_t = Whole depth.
S = Clearance.
C = Center distance.
L = Length of rack.

136

RULES AND FORMULAE FOR SPUR GEAR
CALCULATIONS—Continued

TO FIND	RULE	FORMULA
Diametral pitch P	Divide 3.1416 by the circular pitch.	$P = \dfrac{3.1416}{P_c}$
Circular pitch P_c	Divide 3.1416 by the diametral pitch.	$P_c = \dfrac{3.1416}{P}$
Pitch diameter P_d	Divide the number of teeth by the diametral pitch.	$P_d = \dfrac{N}{P}$
Outside diameter D_o	Add 2 to the number of teeth and divide the sum by the diametral pitch.	$D_o = \dfrac{N + 2}{P}$
Number of teeth N	Multiply the pitch diameter by the diametral pitch.	$N = P_d\,P$
Tooth thickness T	Divide 1.5708 by the diametral pitch.	$T = \dfrac{1.5708}{P}$
Addendum a	Divide 1.0 by the diametral pitch.	$a = \dfrac{1.0}{P}$
Dedendum b	Divide 1.157 by the diametral pitch.	$b = \dfrac{1.157}{P}$
Working depth h_k	Divide 2 by the diametral pitch.	$h_k = \dfrac{2}{P}$
Whole depth h_t	Divide 2.157 by the diametral pitch.	$h_t = \dfrac{2.157}{P}$
Clearance S	Divide 0.157 by the diametral pitch.	$S = \dfrac{0.157}{P}$
Center distance C	Add the number of teeth in both gears and divide the sum by two times the diametral pitch.	$C = \dfrac{N_1 + N_2}{2P}$
Length of rack L	Multiply the number of teeth in the rack by the circular pitch.	$L = N\,P_c$

Sec. VI
Gears

CONVERSION TO METRIC GEAR DESIGN

A large number of the gears used in the manufactured products of today are produced by gear specialists. Metric gears are now part of this production. Gears produced by the American gear industry, using the inch system diametral pitch standard, are not compatible with metric gears which use the module standard. Therefore, a knowledge of the relationship of the two systems is very important. In addition to the different length measurements (inches and millimeters), each system has set up a different standard for its gear tooth proportions. Thus, gears made under these two different standards are not interchangeable.

To understand the basic difference between the inch system (diametral pitch), and the metric system (module), note the following definitions.

DIAMETRAL PITCH: In the inch system, diametral pitch is defined as the number of teeth for each inch of pitch diameter. The diametral pitch is equal to the number of teeth divided by the pitch diameter.

MODULE: In the metric system, module is defined as the amount of pitch diameter per tooth expressed in millimeters. The module is equal to the pitch diameter divided by the number of teeth or the circular pitch divided by 3.1416.

Now the relationship of the two systems can be seen if we compare the amount of pitch diameter per tooth for each. In the inch system the reciprocal of the diametral pitch represents the amount of pitch diameter in inches for each tooth. When this amount is converted to millimeters, it becomes the module or the amount of pitch diameter in millimeters for each tooth.

EXAMPLE: A gear having 40 teeth and a pitch diameter of 5 inches would have a diametral pitch of 8. The reciprocal of 8 expressed in inches is 1/8 inch. Converted to millimeters 1/8 inch is 3.175 millimeters. 3.175 is the module of an equivalent-sized metric gear.

In practice each system has selected preferred sizes. See chart "Diametral Pitch/Module Gear Equivalents" pages 140 and 141. Here you will see that a 3-module gear is the nearest standard gear to an 8 diametral pitch standard gear.

When converting from diametral pitch inch dimensioned gears to metric module millimeter dimensioned gears, or the reverse, use the following rules:

MODULE is equal to 25.4 divided by the diametral pitch.

DIAMETRAL PITCH is equal to 25.4 divided by the module.

RULES AND FORMULAS FOR SPUR GEAR CALCULATIONS—METRIC MODULE STANDARD
(All dimensions in millimeters)

TO FIND	RULE	FORMULA
Module m	Divide circular pitch by 3.1416.	$m = \dfrac{Pc}{3.1416}$
Circular pitch Pc	Multiply module by 3.1416.	$Pc = m \times 3.1416$
Pitch diameter Pd	Multiply module by number of teeth.	$Pd = m \times N$
Outside diameter Do	Add to pitch diameter two times module.	$Do = Pd + 2m$
Root diameter Dr	Subtract from the pitch diameter 2.5 times module.	$Dr = Pd - 2.5m$
Number of teeth N	Divide the pitch diameter by the module.	$N = \dfrac{Pd}{m}$
Tooth thickness T	Multiply 1.5708 times module.	$T = 1.5708m$
Addendum a	Equal to module.	$a = m$
Dedendum b	Equal to 1.25 times module.	$b = 1.25m$
Working depth hk	Equal to 2 times module.	$hk = 2m$
Whole depth ht	Equal to 2.25 times module.	$ht = 2.25m$
Clearance S	Equal to .25 times module.	$S = .25m$
Center distance C	Add the number of teeth in both gears, multiply the sum by the module, and divide this result by two.	$C = \dfrac{m(N_1 + N_2)}{2}$
Length of rack L	Multiply the number of teeth in the rack by the circular pitch.	$L = NPc$

Sec. VI
Gears

Note: See spur gear notation page 136.

DIAMETRAL PITCH/MODULE GEAR EQUIVALENTS

Diametral Pitch	Module	Circular Pitch		Tooth Thickness		Addendum		Whole Depth of Tooth	
		Inches	Millimeters	Inches	Millimeters	Inches	Millimeters	Inches	Millimeters
1		3.1416	79.796	1.5708	39.898	1.0000	25.400	2.1570	54.788
	20	2.4756	62.832	1.2368	31.416	.7874	20.000	1.6752	42.500
1.5		2.0944	55.198	1.0472	26.599	.6667	16.933	1.4380	36.525
	16	1.9789	50.265	.9894	25.133	.6299	16.000	1.3385	34.000
2		1.5708	39.898	.7854	19.949	.5000	12.700	1.0785	27.394
	12	1.4842	37.699	.7421	18.850	.4724	12.000	1.0039	25.500
2.5		1.2566	31.919	.6285	15.959	.4000	10.160	.8628	21.915
	10	1.2368	31.416	.6184	15.708	.3937	10.000	.8566	21.250
3		1.0472	26.599	.5256	13.299	.3335	8.467	.7190	18.265
	8	.9895	25.133	.4947	12.566	.3150	8.000	.6695	17.000
4		.7854	19.949	.3927	9.975	.2500	6.550	.5593	15.697
	6	.7421	18.850	.3711	9.425	.2362	6.000	.5020	12.750

5	5	.6283	15.959	.3142	7.980	.2000	5.080	.4314	10.958
		.6184	15.708	.3092	7.854	.1969	5.000	.4185	10.625
6	4	.5256	13.299	.2618	6.650	.1667	4.233	.3595	9.131
		.4947	12.566	.2474	6.283	.1575	4.000	.3346	8.500
8	3	.3927	9.975	.1963	4.987	.1250	3.175	.2696	6.848
		.3711	9.425	.1855	4.712	.1181	3.000	.2510	6.375
10	2.5	.3142	7.980	.1571	3.990	.1000	2.540	.2157	5.479
		.3092	7.854	.1546	3.927	.0984	2.500	.2092	5.313
12	2	.2618	6.650	.1309	3.325	.0833	2.117	.1798	4.566
		.2474	6.283	.1257	3.142	.0787	2.000	.1673	4.250
16	1.5	.1963	4.987	.0982	2.494	.0625	1.588	.1348	3.424
		.1855	4.712	.0928	2.356	.0591	1.500	.1255	3.188
20	1.25	.1571	3.990	.0785	1.995	.0500	1.270	.1079	2.739
		.1546	3.927	.0775	1.963	.0492	1.250	.1046	2.656
24	1	.1309	3.325	.0654	1.662	.0417	1.058	.0899	2.283
		.1237	3.142	.0618	1.571	.0394	1.000	.0837	2.125

Sec. VI
Gears

RULES AND FORMULAE FOR BEVEL GEAR CALCULATIONS

(Shafts at Right Angles)

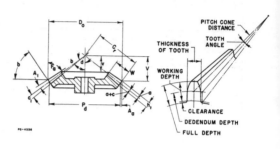

The following symbols are used in conjunction with the formulae for determining the proportions of bevel gear teeth.

Sec. VI
Gears

P = Diametral pitch.

P_c = Circular pitch.

P_d = Pitch diameter.

b = Pitch angle.

C_r = Pitch cone distance.

a = Addendum.

A_1 = Addendum angle.

A_a = Angular addendum.

D_o = Outside diameter.

c_i = Dedendum angle.

$a+c$ = Addendum plus clearance.

a_s = Addendum of small end of tooth.

T_L = Thickness of tooth at pitch line.

T_s = Thickness of tooth at pitch line at small end of gear.

F_a = Face angle.

h_t = Whole depth of tooth space.

V = Apex distance at large end of tooth.

v = Apex distance at small end of tooth.

m_g = Gear ratio.

N = Number of teeth.

N_g = Number of teeth in gear.

N_p = Number of teeth in pinion.

d = Root or cutting angle.

W = Width of gear tooth face.

N_c = Number of teeth of imaginary spur gear for which cutter is selected.

TO FIND	RULE	FORMULA
Diametral pitch P	Divide the number of teeth by the pitch diameter.	$P = \dfrac{N}{P_d}$
Circular pitch P_c	Divide 3.1416 by the diametral pitch.	$P_c = \dfrac{3.1416}{P}$
Pitch diameter P_d	Divide the number of teeth by the diametral pitch.	$P_d = \dfrac{N}{P}$
Pitch angle of pinion $\tan b_p$	Divide the number of teeth in the pinion by the number of teeth in the gear to obtain the tangent.	$\tan b_p = \dfrac{N_p}{N_g}$
Pitch angle of gear $\tan b_g$	Divide the number of teeth in the gear by the number of teeth in the pinion to obtain the tangent.	$\tan b_g = \dfrac{N_g}{N_p}$
Pitch cone distance C_r	Divide the pitch diameter by twice the sine of the pitch angle.	$C_r = \dfrac{P_d}{2 \, (\sin b)}$
Addendum a	Divide 1.0 by the diametral pitch.	$a = \dfrac{1.0}{P}$
Addendum angle $\tan A_1$	Divide the addendum by the pitch cone distance to obtain the tangent.	$\tan A_1 = \dfrac{a}{C_r}$
Angular addendum A_a	Multiply the addendum by the cosine of the pitch angle.	$A_a = a \cos b$
Outside diameter D_o	Add twice the angular addendum to the pitch diameter.	$D_o = P_d + 2A_a$
Dedendum angle $\tan c_1$	Divide the dedendum by the pitch cone distance to obtain the tangent.	$\tan c_1 = \dfrac{a+c}{C_r}$
Addendum of small end of tooth a_s	Subtract the width of face from the pitch cone distance, divide the remainder by the pitch cone distance and multiply by the addendum.	$a_s = a \left(\dfrac{C_r - W}{C_r} \right)$
Thickness of tooth at pitch line T_L	Divide the circular pitch by 2.	$T_L = \dfrac{P_c}{2}$
Thickness of tooth at pitch line at small end of gear T_s	Subtract the width of face from the pitch cone distance, divide the remainder by the pitch cone distance and multiply by the thickness of tooth at the pitch line.	$T_s = T_L \left(\dfrac{C_r - W}{C_r} \right)$
Face angle F_a	Face cone of blank turned parallel to root cone of mating gear.	$F_a = b + c_1$
Whole depth of tooth space h_t	Divide 2.157 by the diametral pitch.	$h_t = \dfrac{2.157}{P}$
Apex distance at large end of tooth V	Multiply one-half the outside diameter by the cotangent of the face angle.	$V = \left(\dfrac{D_o}{2} \right) \cot F_a$
Apex distance at small end of tooth v	Subtract the width of face from the pitch cone distance, divide the remainder by the pitch cone distance and multiply by the apex distance.	$v = V \left(\dfrac{C_r - W}{C_r} \right)$
Gear ratio m_g	Divide the number of teeth in the gear by the number of teeth in the pinion.	$m_g = \dfrac{N_g}{N_p}$
Number of teeth in gear and/or pinion N_g, N_p	Multiply the pitch diameter by the diametral pitch.	$N_g = P_d P$ $N_p = P_d P$
Cutting angle d	Subtract the dedendum angle from the pitch angle.	$d = b - c_1$
Number of teeth of imaginary spur gear for which cutter is selected N_o	Divide the number of teeth in actual gear by the cosine of the pitch angle.	$N_o = \dfrac{N}{\cos b}$

Sec. VI
Gears

RULES AND FORMULAE FOR WORM GEAR
CALCULATIONS (Solid Type)
(Single and Double Thread—14½° Pressure Angle)

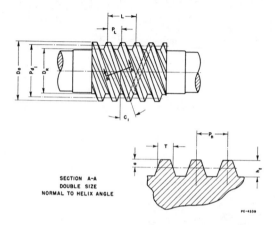

SECTION A-A
DOUBLE SIZE
NORMAL TO HELIX ANGLE

PC-4339

The following symbols are used in conjunction with the formulae for determining the proportions of worm gear teeth.

P_L = Linear pitch.
P_{d_1} = Pitch diameter.
D_o = Outside diameter.
N_w = Number of threads.
D_R = Root diameter.
h_t = Whole depth of tooth.
C_1 = Helix angle.
P_n = Normal pitch.
a = Addendum.
L = Lead.
T = Normal tooth thickness.
t = Width of thread tool at end.

RULES AND FORMULAE FOR WORM GEAR
CALCULATIONS (Solid Type)—Continued

(Single and Double Thread—$14\frac{1}{2}°$ Pressure Angle)

TO FIND	RULE	FORMULA
Linear pitch P_L	Divide the lead by the number of threads in the whole worm: i. e., one if single-threaded or four if quadrupled threaded.	$P_L = \dfrac{L}{N_w}$
Pitch diameter P_{d_1}	Subtract twice the addendum from the out-side diameter.	$P_{d_1} = D_o - 2\,a$
Outside diameter D_o	Add twice the addendum of the worm to the pitch diameter of the worm	$D_o = P_{d_1} + 2\,a$
Root diameter D_R	Subtract twice the whole depth of the tooth from the outside diameter.	$D_R = D_o - 2\,h_t$
Whole depth of tooth h_t	Multiply the linear pitch by 0.6866.	$h_t = 0.6866 P_L$
Helix angle C_1	Multiply the pitch diameter of the worm by 3.1416, and divide the product by the lead. The quotient is the cotangent of the helix angle.	$\cot C_1 = \dfrac{3.1416 P_{d_1}}{L}$
Normal pitch P_n	Multiply the linear pitch by the cosine of the helix angle of the worm.	$P_n = P_L \cos C_1$
Addendum a	Multiply the linear pitch by 0.3183.	$a = 0.3183\,P_L$
Lead L	Multiply the linear pitch by the number of threads.	$L = P_L N_w$
Normal tooth thickness T	Multiply one-half the linear pitch by the cosine of the helix angle.	$T = \dfrac{P_L}{2} \cos C_1$
Width of thread tool at end t	Multiply the linear pitch by 0.31.	$t = 0.31\,P_L$

RULES AND FORMULAE FOR WORM WHEEL CALCULATIONS

(Single and Double Thread—$14\frac{1}{2}°$ Pressure Angle)

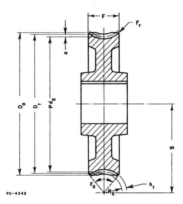

PC-4342

The following symbols are used in conjunction with the formulae for determining the proportions of worm wheel teeth.

P_c = Circular pitch.

P_{d_1} = Pitch diameter.

N = Number of teeth.

D_o = Outside diameter.

D_t = Throat diameter.

R_c = Radius of curvature of worm wheel throat.

D = Diameter to sharp corners.

F_a = Face angle.

F = Face width of rim.

F_r = Radius at edge of face.

a = Addendum.

h_t = Whole depth of tooth.

S = Center distance between worm and worm wheel.

G = Gashing angle.

TO FIND	RULE	FORMULA
Circular pitch P_c	Divide the pitch diameter by the product of 0.3183 and the number of teeth.	$P_c = \dfrac{P_{d_1}}{0.3183\ N}$
Pitch diameter P_{d_1}	Multiply the number of teeth in the worm wheel by the linear pitch of the worm, and divide the product by 3.1416.	$P_{d_1} = \dfrac{N P_L}{3.1416}$
Outside diameter D_o	Multiply the circular pitch by 0.4775 and add the product to the throat diameter.	$D_o = D_t + 0.4775\ P_c$
Throat diameter D_t	Add twice the addendum of the worm tooth to the pitch diameter of the worm wheel.	$D_t = P_{d_1} + 2\ a$
Radius of curvature of worm wheel throat R_c	Subtract twice the addendum of the worm tooth from half the outside diameter of the worm.	$R_c = \dfrac{D_o}{2} - 2\ a$
Diameter to sharp corners D	Multiply the radius of curvature of the worm-wheel throat by the cosine of half the face angle, subtract this quantity from the radius of curvature. Multiply the remainder by 2, and add the product to the throat diameter of the worm wheel.	$D = 2\left(R_c - R_c \times \cos\dfrac{F_a}{2}\right) + D_t$
Face width of rim F	Multiply the circular pitch by 2.38 and add 0.25 to the product.	$F = 2.38\ P_c + 0.25$
Radius at edge of face F_r	Divide the circular pitch by 4.	$F_r = \dfrac{P_c}{4}$
Addendum a	Multiply the circular pitch by 0.3183.	$a = 0.3183\ P_c$
Whole depth of tooth h_t	Multiply the circular pitch by 0.6866.	$h_t = 0.6866\ P_c$
Center distance between worm and worm wheel S	Add the pitch diameter of the worm to the pitch diameter of the worm wheel and divide the sum by 2.	$S = \dfrac{P_{d_1} + P_{d_1}}{2}$
Gashing angle G	Divide the lead of the worm by the circumference of the pitch circle. The result will be the cotangent of the gashing angle.	$\cot G = \dfrac{L}{3.1416\ d}$

Sec. VI
Gears

RULES AND FORMULAE FOR HELICAL GEAR CALCULATIONS

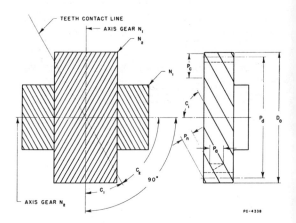

The following symbols are used in conjunction with the formulae for determining the proportions of helical gear teeth.

P_{nd} = Normal diametral pitch (pitch of cutter).
P_c = Circular pitch.
P_a = Axial pitch.
P_n = Normal pitch.
P_d = Pitch diameter.
S = Center distance.
C, C_1, C_2 = Helix angle of the gears.
L = Lead of tooth helix.
T_n = Normal tooth thickness at pitch line.
a = Addendum.
h_t = Whole depth of tooth.
N, N_1, N_2 = Number of teeth in the gears.
D_o = Outside diameter.
N_c = Hypothetical number of teeth for which the gear cutter should be selected.

TO FIND	RULE	FORMULA
Normal diametral pitch P_{nd}	Divide the number of teeth by the product of the pitch diameter and the cosine of the helix angle.	$P_{nd} = \dfrac{N}{P_d \cos C_1}$
Circular pitch P_c	Multiply the pitch diameter of the gear by 3.1416, and divide the product by the number of teeth in the gear.	$P_c = \dfrac{3.1416\, P_d}{N}$
Axial pitch P_a	Multiply the circular pitch by the cotangent of the helix angle.	$P_a = P_c \cot C_1$
Normal pitch P_n	Divide 3.1416 by the normal diametral pitch.	$P_n = \dfrac{3.1416}{P_{nd}}$
Pitch diameter P_d	Divide the number of teeth by the product of the normal pitch and the cosine of the helix angle.	$P_d = \dfrac{N}{P_{nd} \cos C_1}$
Center distance S	Divide the sum of the pitch diameters of the mating gears by 2.	$S = \dfrac{P_{d_1} + P_{d_1}}{2}$
Checking Formulae (shafts at right angles)	Multiply the number of teeth in the first gear by the tangent of the tooth angle of that gear, and add the number of teeth in the second gear to the product. The sum should equal twice the product of the center distance multiplied by the normal diametral pitch, multiplied by the sine of the helix angle.	$N_1 + (N_2 \tan C_2) = 2\, S\, P_{nd} \sin C_1$
Lead of tooth helix L	Multiply the pitch diameter by 3.1416 times the cotangent of the helix angle.	$L = 3.1416\, P_d \cot C_1$
Normal tooth thickness at pitch line T_n	Divide 1.571 by the normal diametral pitch.	$T_n = \dfrac{1.571}{P_{nd}}$
Addendum a	Divide the normal pitch by 3.1416.	$a = \dfrac{P_n}{3.1416}$
Whole depth of tooth h_t	Divide 2.157 by the normal diametral pitch.	$h_t = \dfrac{2.157}{P_{nd}}$
Outside diameter D_o	Add twice the addendum to the pitch diameter.	$D_o = P_d + 2\,a$
Hypothetical number of teeth for which gear cutter should be selected N_c	Divide the number of teeth in the gear by the cube of the cosine of the helix angle.	$N_c = \dfrac{N_1}{(\cos C_1)^3}$

Sec. VI
Gears

Weights of ● and ■ Steel Lineal Foot
Based on 489.6 Lbs. per Cubic Foot

Size Inches	Weight of ● 1-foot Long	Weight of ■ 1-foot Long	Size Inches	Weight of ● 1-foot Long	Weight of ■ 1-foot Long
1/16	.0104	.0133	3 1/16	25.04	31.89
1/8	.0417	.0531	3 1/8	26.08	33.20
3/16	.0938	.1195	3 3/16	27.13	34.55
1/4	.1669	.2123	3 1/4	28.20	35.92
5/16	.2608	.3333	3 5/16	29.30	37.31
3/8	.3756	.4782	3 3/8	30.42	38.73
7/16	.5111	.6508	3 7/16	31.56	40.18
1/2	.6676	.8500	3 1/2	32.71	41.65
9/16	.8449	1.076	3 9/16	33.90	43.14
5/8	1.043	1.328	3 5/8	35.09	44.68
11/16	1.262	1.608	3 11/16	36.31	46.24
3/4	1.502	1.913	3 3/4	37.56	47.82
13/16	1.763	2.245	3 13/16	38.81	49.42
7/8	2.044	2.603	3 7/8	40.10	51.05
15/16	2.347	2.989	3 15/16	41.40	52.71
1	2.670	3.400	4	42.73	54.40
1 1/16	3.014	3.838	4 1/16	44.07	56.11
1 1/8	3.379	4.303	4 1/8	45.44	57.85
1 3/16	3.766	4.795	4 3/16	46.83	59.62
1 1/4	4.173	5.312	4 1/4	48.24	61.41
1 5/16	4.600	5.857	4 5/16	49.66	63.23
1 3/8	5.019	6.428	4 3/8	51.11	65.08
1 7/16	5.518	7.026	4 7/16	52.58	66.95
1 1/2	6.008	7.650	4 1/2	54.07	68.85
1 9/16	6.520	8.301	4 9/16	55.59	70.78
1 5/8	7.051	8.978	4 5/8	57.12	73.73
1 11/16	7.604	9.682	4 11/16	58.67	74.70
1 3/4	8.178	10.41	4 3/4	60.25	76.71
1 13/16	8.773	11.17	4 13/16	61.84	78.74
1 7/8	9.388	11.95	4 7/8	63.46	80.81
1 15/16	10.02	12.76	4 15/16	65.10	82.89
2	10.68	13.60	5	66.76	85.00
2 1/16	11.36	14.46	5 1/16	68.44	87.14
2 1/8	12.06	15.35	5 1/8	70.14	89.30
2 3/16	12.78	16.27	5 3/16	71.86	91.49
2 1/4	13.52	17.22	5 1/4	73.60	93.72
2 5/16	14.28	18.19	5 5/16	75.37	95.96
2 3/8	15.07	19.18	5 3/8	77.15	98.23
2 7/16	15.86	20.02	5 7/16	78.95	100.5
2 1/2	16.69	21.25	5 1/2	80.77	102.8
2 9/16	17.53	22.33	5 9/16	82.62	105.2
2 5/8	18.40	23.43	5 5/8	84.49	107.6
2 11/16	19.29	24.56	5 11/16	86.38	110.0
2 3/4	20.20	25.00	5 3/4	88.29	112.4
2 13/16	21.12	26.90	5 13/16	90.22	114.9
2 7/8	22.07	28.10	5 7/8	92.17	117.4
3	24.03	30.60	6	96.14	122.4

U. S. STANDARD SHEET METAL GAGE NUMBERS
WITH APPROXIMATE THICKNESSES AND WEIGHTS

GAGE NUMBER	APPROX. THICKNESS IN DECIMALS	WEIGHT PER SQ. FT. IN OUNCES	WEIGHT PER SQ. FT. IN POUNDS
0000000	.5000	320	20.00
000000	.4688	300	18.75
00000	.4375	280	17.50
0000	.4063	260	16.25
000	.3750	240	15.00
00	.3438	220	13.75
0	.3125	200	12.50
1	.2813	180	11.25
2	.2656	170	10.62
3	.2500	160	10.00
4	.2344	150	9.37
5	.2188	140	8.75
6	.2031	130	8.12
7	.1875	120	7.50
8	.1719	110	6.87
9	.1563	100	6.25
10	.1406	90	5.62
11	.1250	80	5.00
12	.1094	70	4.37
13	.0938	60	3.75
14	.0781	50	3.12
15	.0703	45	2.81
16	.0625	40	2.50
17	.0563	36	2.25
18	.0500	32	2.00
19	.0438	28	1.75
20	.0375	24	1.500
21	.0344	22	1.375
22	.0313	20	1.250
23	.0281	18	1.125
24	.0250	16	1.000
25	.0219	14	0.875
26	.0188	12	0.750
27	.0172	11	0.688
28	.0156	10	0.625
29	.0141	9	0.563
30	.0125	8	0.500
31	.0109	7	0.438
32	.0102	6½	0.406
33	.0094	6	0.375
34	.0086	5½	0.344
35	.0078	5	0.313
36	.0070	4½	0.281
37	.0066	4¼	0.266
38	.0063	4	0.250
39	.0059	3¾	0.234
40	.0055	3½	0.219
41	.0053	3⅜	0.211
42	.0051	3¼	0.203
43	.0049	3⅛	0.195
44	.0047	3	0.188

Sec. VII
Weights,
Gages,
Tolerances

STANDARD WIRE GAUGES

Number of Wire Gauge	Birmingham or Stubs, Inches	American or Brown & Sharpe, Inches	Washburn & Moen Mfg. Co., Inches	Weight per 100 Feet W. & M. Gauge, Pounds	Music Wire Gauge, Inches
0000	.454	.460	.393	40.94	....
000	.425	.410	.362	34.73	....
00	.380	.365	.331	29.04	.0085
0	.340	.325	.307	27.66	.009
1	.300	.289	.283	21.23	.010
2	.284	.258	.263	18.34	.011
3	.259	.229	.244	15.78	.012
4	.238	.204	.225	13.39	.013
5	.220	.182	.207	11.35	.014
6	.203	.162	.192	9.73	.016
7	.180	.144	.177	8.03	.018
8	.165	.128	.162	6.96	.020
9	.148	.114	.148	5.08	.022
10	.134	.102	.135	4.83	.024
11	.120	.091	.120	3.82	.026
12	.109	.081	.105	2.92	.028
13	.095	.072	.092	2.24	.030
14	.083	.064	.080	1.69	.032
15	.072	.057	.072	1.37	.034
16	.065	.051	.063	1.05	.036
17	.058	.045	.054	.77	.038
18	.049	.040	.047	.58	.040
19	.042	.036	.041	.45	.042
20	.035	.032	.035	.32	.044
21	.032	.028	.032	.27	.046
22	.028	.025	.028	.21	.048
23	.025	.023	.025	.175	.051
24	.022	.020	.023	.140	.055
25	.020	.018	.020	.116	.059
26	.018	.016	.018	.093	.063
27	.016	.014	.017	.083	.067
28	.014	.0125	.016	.074	.071
29	.013	.011	.015	.061	.074
30	.012	.010	.014	.054	.078
31	.010	.009	.0135	.050	.082
32	.009	.008	.013	.046	.086
33	.008	.007	.011	.037	.090
34	.007	.0063	.010	.030	....
35	.005	.0056	.0095	.025	....
36	.004	.005	.009	.021	....

Standard Keyways and Collar Diameters

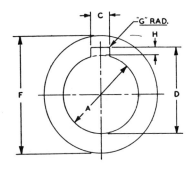

Cutter Bore "A" Inches	Nominal Size Key (Square) Inches	Dimensions—Inches						
		C		D		H	G Corner Radius	F
		Max.	Min.	Max.	Min.	Nominal		Collar
½	3/32	.106	.099	.5678	.5578	3/64	.020	7/8
5/8	1/8	.137	.130	.7085	.6985	1/16	1/32	1
3/4	1/8	.137	.130	.8325	.8225	1/16	1/32	1 1/8
7/8	1/8	.137	.130	.9575	.9475	1/16	1/32	1 1/4
1	1/4	.262	.255	1.1140	1.1040	3/32	3/64	1 1/2
1 1/4	5/16	.325	.318	1.3950	1.3850	1/8	1/16	1 7/8
1 1/2	3/8	.410	.385	1.6760	1.6660	5/32	1/16	2 1/4
1 3/4	7/16	.473	.448	1.9580	1.9480	3/16	1/16	2 1/2
2	1/2	.535	.510	2.2080	2.1980	3/16	1/16	2 3/4
2 1/2	5/8	.660	.635	2.7430	2.7330	7/32	1/16	3 1/2
3	3/4	.785	.760	3.2750	3.2650	1/4	3/32	4
3 1/2	7/8	.910	.885	3.9000	3.8900	3/8	3/32	5
4	1	1.035	1.011	4.4000	4.3900	3/8	3/32	5 1/2
4 1/2	1 1/8	1.160	1.135	4.9630	4.9530	7/16	1/8	6
5	1 1/4	1.285	1.260	5.5250	5.5150	1/2	1/8	6 1/2

For intermediate size bore use the Keyway for the next larger size bore listed.

Woodruff Keys

No.	S.A.E. No.	Amer. Std. No.	Size Inches
201	...	...	1/4 x 1/16
206	...	...	5/16 x 1/16
207	...	...	5/16 x 3/32
211	...	...	3/8 x 1/16
212	...	...	3/8 x 3/32
213	...	...	3/8 x 1/8
1	10	204	1/2 x 1/16
2	20	304	1/2 x 3/32
3	30	404	1/2 x 1/8
4	40	305	5/8 x 3/32
5	50	405	5/8 x 1/8
6	60	505	5/8 x 5/32
61	...	...	5/8 x 3/16
7	70	406	3/4 x 1/8
8	80	506	3/4 x 5/32
9	90	606	3/4 x 3/16
91	...	...	3/4 x 1/4
10	100	507	7/8 x 5/32
11	110	607	7/8 x 3/16
12	...	...	7/8 x 7/32
A	115	807	7/8 x 1/4
121	...	...	7/8 x 1/4
13	130	608	1 x 3/16
14	...	...	1 x 7/32
15	150	808	1 x 1/4
B	155	1008	1 x 5/16
131	...	...	1 x 5/16
152	...	...	1 x 3/8
141	...	...	1 x 1/4
16	160	609	1 1/8 x 3/16
17	...	...	1 1/8 x 7/32
18	180	809	1 1/8 x 1/4
C	185	1009	1 1/8 x 5/16
161	...	...	1 1/8 x 5/16
19	...	...	1 1/4 x 3/16
20	...	...	1 1/4 x 7/32
21	210	810	1 1/4 x 1/4
D	215	1010	1 1/4 x 5/16
E	225	1210	1 1/4 x 3/8
22	...	811	1 3/8 x 1/4
23	...	1011	1 3/8 x 5/16
F	...	1211	1 3/8 x 3/8
24	...	812	1 1/2 x 1/4
25	...	1012	1 1/2 x 5/16
G	...	1212	1 1/2 x 3/8
126	...	...	2 1/8 x 3/16
127	...	...	2 1/8 x 1/4
128	...	...	2 1/8 x 5/16
129	...	...	2 1/8 x 3/8
26	...	...	2 1/8 x 3/16
27	...	...	2 1/8 x 1/4
28	...	...	2 1/8 x 5/16
29	...	...	2 1/8 x 3/8
Rx	...	...	2 3/4 x 1/4
Sx	...	...	2 3/4 x 5/16
Tx	...	...	2 3/4 x 3/8
Ux	...	...	2 3/4 x 7/16
Vx	...	...	2 3/4 x 1/2
R	...	...	2 3/4 x 1/4
S	...	...	2 3/4 x 5/16
T	...	...	2 3/4 x 3/8
U	...	...	2 3/4 x 7/16
V	...	...	2 3/4 x 1/2
30	...	...	3 1/2 x 3/8
31	...	...	3 1/2 x 7/16
32	...	...	3 1/2 x 1/2
33	...	...	3 1/2 x 9/16
34	...	...	3 1/2 x 5/8
35	...	...	3 1/2 x 11/16
36	...	...	3 1/2 x 3/4

American Standard

Preferred Limits and Fits for Cylindrical Parts

ASA B4.1-1955

Definitions.

Nominal Size. The nominal size is the designation which is used for the purpose of general identification.

Dimension. A dimension is a geometrical characteristic such as diameter, length, angle, or center distance.

Size. Size is a designation of magnitude. When a value is assigned to a dimension it is referred to hereinafter as the size of that dimension.

> NOTE: *It is recognized that the words "dimension" and "size" are both used at times to convey the meaning of magnitude.*

Allowance. An allowance is an intentional difference between the maximum material limits of mating parts. (See definition of "Fit.") It is a minimum clearance (positive allowance) or maximum interference (negative allowance) between mating parts.

Tolerance. A tolerance is the total permissible variation of a size. The tolerance is the difference between the limits of size.

Basic Size. The basic size is that size from which the limits of size are derived by the application of allowances and tolerances.

Design Size. The design size is that size from which the limits of size are derived by the application of tolerances. When there is no allowance the design size is the same as the basic size.

Actual Size. An actual size is a measured size.

Limits of Size. The limits of size are the applicable maximum and minimum sizes.

Maximum Material Limit. A maximum material limit is the maximum limit of size of an external dimension or the minimum limit of size of an internal dimension.

Minimum Material Limit. A minimum material limit is the minimum limit of size of an external dimension or the maximum limit of size of an internal dimension.

Tolerance Limit. A tolerance limit is the variation, positive or negative, by which a size is permitted to depart from the design size.

Unilateral Tolerance. A unilateral tolerance is a tolerance in which variation is permitted only in one direction from the design size.

Bilateral Tolerance. A bilateral tolerance is a tolerance in which variation is permitted in both directions from the design size.

Unilateral Tolerance System. A design plan which uses only unilateral tolerances is known as a Unilateral Tolerance System.

Bilateral Tolerance System. A design plan which uses only bilateral tolerances is known as a Bilateral Tolerance System.

Fit. Fit is the general term used to signify the range of tightness which may result from the application of a specific combination of allowances and tolerances in the design of mating parts.

Actual Fit. The actual fit between two mating parts is the relation existing between them with respect to the amount of clearance or interference which is present when they are assembled.

NOTE: *Fits are of three general types: clearance, transition, and interference.*

Clearance Fit. A clearance fit is one having limits of size so prescribed that a clearance always results when mating parts are assembled.

Interference Fit. An interference fit is one having limits of size so prescribed that an interference always results when mating parts are assembled.

Transition Fit. A transition fit is one having limits of size so prescribed that either a clearance or an interference may result when mating parts are assembled.

Basic Hole System. A basic hole system is a system of fits in which the design size of the hole is the basic size and the allowance is applied to the shaft.

Basic Shaft System. A basic shaft system is a system of fits in which the design size of the shaft is the basic size and the allowance is applied to the hole.

Designation of Standard Fits. Standard fits are designated by means of the symbols given below to facilitate reference to classes of fit for educational purposes. These symbols are not intended to be shown on manufacturing drawings; instead, sizes should be specified on drawings.

The letter symbols used are as follows:

RC   Running or Sliding Fit
LC   Locational Clearance Fit
LT   Transition Fit
LN   Locational Interference Fit
FN   Force or Shrink Fit

These letter symbols are used in conjunction with numbers representing the class of fit; thus "FN 4" represents a class 4, force fit.

Standard Tolerances. The series of standard tolerances shown in the table below are so arranged that for any one grade they represent approximately similar production difficulties throughout the range of sizes. The table provides a suitable range from which appropriate tolerances for holes and shafts can be selected. This enables the use of standard gages.

Tolerance values are in thousandths of an inch. Data in bold face are in accordance with ABC agreements.

Nominal Size Range Inches Over — To	Grade 4	Grade 5	Grade 6	Grade 7	Grade 8	Grade 9	Grade 10	Grade 11	Grade 12	Grade 13
0.04– 0.12	0.15	0.20	0.25	0.4	0.6	1.0	1.6	2.5	4	6
0.12– 0.24	0.15	0.20	0.3	0.5	0.7	1.2	1.8	3.0	5	7
0.24– 0.40	0.15	0.25	0.4	0.6	0.9	1.4	2.2	3.5	6	9
0.40– 0.71	0.2	0.3	0.4	0.7	1.0	1.6	2.8	4.0	7	10
0.71– 1.19	0.25	0.4	0.5	0.8	1.2	2.0	3.5	5.0	8	12
1.19– 1.97	0.3	0.4	0.6	1.0	1.6	2.5	4.0	6	10	16
1.97– 3.15	0.3	0.5	0.7	1.2	1.8	3.0	4.5	7	12	18
3.15– 4.73	0.4	0.7	0.9	1.4	2.2	3.5	5	9	14	22
4.73– 7.09	0.5	0.8	1.0	1.6	2.5	4.0	6	10	16	25
7.09– 9.85	0.6	0.9	1.2	1.8	2.8	4.5	7	12	18	28
9.85– 12.41	0.6	1.0	1.2	2.0	3.0	5.0	8	12	20	30
12.41– 15.75	0.7	1.0	1.4	2.2	3.5	6	9	14	22	35
15.75– 19.69	0.8	1.2	1.6	2.5	4	8	10	16	25	40
19.69– 30.09	0.9	1.6	2.0	3	5	10	12	20	30	50
30.09– 41.49	1.0	2.0	2.5	4	6	12	16	25	40	60
41.49– 56.19	1.2	2.5	3	5	8	16	20	30	50	80
56.19– 76.39	1.6	3	4	6	10	20	25	40	60	100
76.39–100.9	2.0	4	5	8	12	25	30	50	80	125
100.9– 131.9	2.5	5	6	10	16	30	40	80	100	160
131.9– 171.9	3	6	8	12	20		50	100	125	200
171.9– 200	4		10	16	25	40	60		160	250

Machining Processes. To indicate the machining processes which may normally be expected to produce work within the tolerances indicated by the grades given in this Standard, Fig. 1 has been provided. This information is intended merely as a guide.

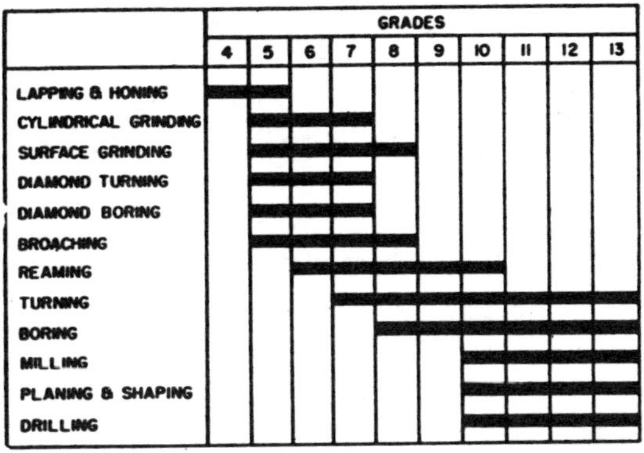

	GRADES									
	4	5	6	7	8	9	10	11	12	13
LAPPING & HONING	■	■								
CYLINDRICAL GRINDING		■	■	■						
SURFACE GRINDING		■	■	■	■					
DIAMOND TURNING		■	■	■						
DIAMOND BORING		■	■	■						
BROACHING		■	■	■	■					
REAMING			■	■	■	■	■			
TURNING				■	■	■	■	■	■	■
BORING					■	■	■	■	■	■
MILLING					■	■	■	■	■	■
PLANING & SHAPING					■	■	■	■	■	■
DRILLING						■	■	■	■	■

Fig. 1 Machining Processes

LIMITS FOR TURNING AND GRINDING

The limits given in the tables, pages 160 and 161, are recommended for use in the manufacture of machine parts, to produce satisfactory commercial work. These limits should only be followed under ordinary conditions. For special cases, it may be necessary to increase or decrease the limits given in the tables. The allowance to be used when rough turning parts to be ground varies from 0.010 to 0.030 inch; that is, a part to be ground to a diameter of 1 inch would be rough turned in the lathe to a diameter of from 1.010 to 1.015 inches, while a 3-inch shaft may have an allowance of from 0.015 to 0.025 inch. The allowance depends largely on the class of work.

Sec. VII
Weights,
Gages,
Tolerances

Allowances for Fits

Grinding Limits for Cylindrical Parts

Diameter, Inches	Limits, Inches
Running Fits -- Ordinary Speed	
Up to 1/2	− 0.00025 to − 0.00075
1/2 to 1	− 0.00075 to − 0.0015
1 to 2	− 0.0015 to − 0.0025
2 to 3-1/2	− 0.0025 to − 0.0035
3-1/2 to 6	− 0.0035 to − 0.005
Running Fits -- High-Speed, Heavy Pressure and Rocker Shafts	
Up to 1/2	− 0.0005 to − 0.001
1/2 to 1	− 0.001 to − 0.002
1 to 2	− 0.002 to − 0.003
2 to 3-1/2	− 0.003 to − 0.0045
3-1/2 to 6	− 0.0045 to − 0.0065
Sliding Fits	
Up to 1/2	− 0.00025 to − 0.0005
1/2 to 1	− 0.0005 to − 0.001
1 to 2	− 0.001 to − 0.002
2 to 3-1/2	− 0.002 to − 0.0035
3-1/2 to 6	− 0.003 to − 0.005

Allowances for Fits

Grinding Limits for Cylindrical Parts

Diameter, Inches	Limits, Inches
Driving Fits -- Ordinary	
Up to 1/2	+ 0.00025 to + 0.0005
1/2 to 1	+ 0.001 to + 0.002
1 to 2	+ 0.002 to + 0.003
2 to 3-1/2	+ 0.003 to + 0.004
3-1/2 to 6	+ 0.004 to + 0.005
Forced Fits	
Up to 1/2	+ 0.00075 to + 0.0015
1/2 to 1	+ 0.0015 to + 0.0025
1 to 2	+ 0.0025 to + 0.004
2 to 3-1/2	+ 0.004 to + 0.006
3-1/2 to 6	+ 0.006 to + 0.009
Driving Fits -- For such Pieces as are Required to be Readily Taken Apart	
Up to 1/2	+ 0 to + 0.00025
1/2 to 1	+ 0.00025 to + 0.0005
1-1/2 to 2	+ 0.0005 to + 0.00075
2 to 3-1/2	+ 0.00075 to + 0.001
3-1/2 to 6	+ 0.001 to + 0.0015

SELECTING CLASS OF FIT

The machine designer in most cases is responsible for selecting the basic sizes, limits of size, allowances, and tolerances when specifying the dimensions of parts for machinery to be used in interchangeable manufacture. It is the responsibility of those machining the parts to produce them within the prescribed limits. The choice of these dimensions is complicated by the factors involved and all of them must be considered when selecting the proper grade or class of fit. A look at the chart on page 158 will show this wide variation of tolerances. A guide for the selection of the proper machining process to be used for producing the parts within tolerance is suggested in Fig. 1, page 159.

One prominent manufacturer has recommended the following "Allowances for Fits" as shown on pages 160 and 161, where the plus (+) designates larger than basic size; and minus (−) designates smaller than basic size.

Charts covering the complete range of grades for the different classes of fits as recommended and established by the American, British, and Canadian Conference (A.B.C.) are available in bulletin ASA B 4.1-1955 published by the American Society of Mechanical Engineers, New York, New York 10017. The information on pages 155-158 was extracted from pages 5-8 and 17 of the bulletin with the publisher's permission.

SINE BAR or SINE PLATE SETTING

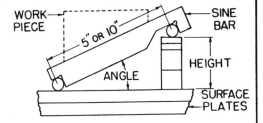

Sine bars or sine plates usually have a length of 5" or 10". These standard lengths are commonly used by the tool maker or inspector. The sine bar or sine plate is used for accurately setting up work for machining or for inspection. Gage blocks are usually used for establishing the height, as shown in the figure above. Refer to page 166 for information on the selection of the proper gage blocks to establish a given dimension.

Rule for determining the height of the sine bar setting for a given angle: multiply the sine of the angle by the length of the sine bar. The sine of the angle is taken from the tables of trigonometric functions starting on page 29.

Problem: What would be the height to set a sine bar for establishing an angle of 23°41'? Solution: The sine of 23°41' is .40168. Multiply this by 5 because a 5" sine bar is used; 5 X .40168 = 2.0084, which is the height to set the sine bar.

PIN MEASUREMENTS FOR GIBS AND DOVETAILS

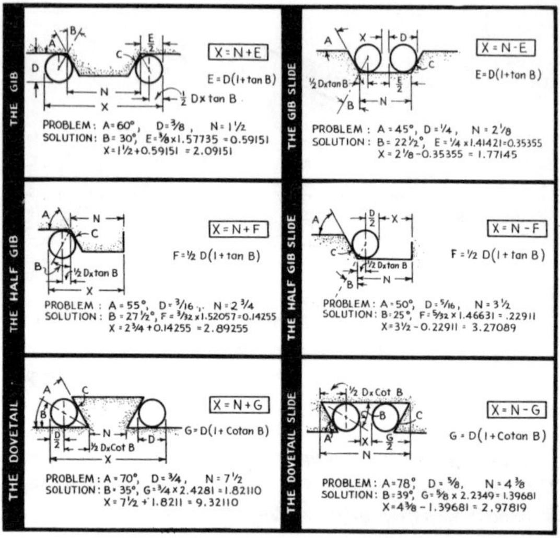
THE GIB
X=N+E
E=D(1+tan B)
½ Dx tan B
PROBLEM: A=60°, D=3/8, N=1½
SOLUTION: B=30°, E=3/8×1.57735=0.59151
X=1½+0.59151=2.09151
THE GIB SLIDE
X=N-E
E=D(1+tan B)
½ DxtanB
PROBLEM: A=45°, D=¼, N=2⅛
SOLUTION: B=22½°, E=¼×1.41421=0.35355
X=2⅛-0.35355=1.77145
THE HALF GIB
X=N+F
F=½ D(1+tan B)
½ Dxtan B
PROBLEM: A=55°, D=3/16, N=2¾
SOLUTION: B=27½°, F=3/32×1.52057=0.14255
X=2¾+0.14255=2.89255
THE HALF GIB SLIDE
X=N-F
F=½ D(1+tan B)
½ Dxtan B
PROBLEM: A=50°, D=5/16, N=3½
SOLUTION: B=25°, F=5/32×1.46631=.22911
X=3½-0.22911=3.27089
THE DOVETAIL
X=N+G
G=D(1+Cotan B)
½ DxCot B
PROBLEM: A=70°, D=¾, N=7½
SOLUTION: B=35°, G=¾×2.4281=1.82110
X=7½+1.8211=9.32110
THE DOVETAIL SLIDE
X=N-G
G=D(1+Cotan B)
½ DxCot B
PROBLEM: A=78°, D=5/8, N=4⅜
SOLUTION: B=39°, G=5/8×2.2349=1.39681
X=4⅜-1.39681=2.97819

CONSTANTS FOR CALCULATING ABOVE CASES WITH COMMON ANGLES AND STANDARD PINS

STD PIN. DIA.		E		F		G	
Fraction	Decimal	A=45°	60°	A=45°	60°	A=45°	60°
3/64	0.04688	0.06629	0.07394	0.03315	0.03690	0.16004	0.12806
1/16.	0.06 25	0.08839	0.09858	0.04419	0.04929	0.21339	0.17075
5/64	0.07813	0.11048	0.12323	0.05524	0.06162	0.26673	0.21344
3/32	0.09375	0.13258	0.14788	0.06629	0.07381	0.32082	0.25613
1/8	0.1250	0.17678	0.19717	0.08839	0.09858	0.42677	0.34150
5/32	0.15625	0.22097	0.24646	0.11048	0.12323	0.53347	0.42688
3/16	0.18750	0.26516	0.29575	0.13258	0.14763	0.64016	0.51226
1/4	0.250	0.35355	0.39434	0.17678	0.19717	0.85355	0.68301
5/16	0.3125	0.44194	0.49292	0.22097	0.24646	1.06694	0.85376
3/8	0.375	0.53033	0.59151	0.26516	0.29575	1.28033	1.02452
7/16	0.4375	0.61872	0.69009	0.30936	0.34504	1.49372	1.19527
1/2	0.500	0.70710	0.78867	0.35355	0.39434	1.70710	1.36602
5/8	0.625	0.88388	0.98584	0.44194	0.49292	2.13388	1.70753
3/4	0.750	1.06066	1.18301	0.53033	0.59151	2.56066	2.04904
1"	1.000	1.41421	1.57735	0.70711	0.78867	3.41421	2.73205
1¼	1.250	1.76776	1.97168	0.88388	0.98584	4.26776	3.41406
1½	1.500	2.12131	2.36602	1.06066	1.18301	5.12132	4.09808
2"	2.000	2.82842	3.15470	1.41421	1.57735	6.82842	5.46410

PIN MEASUREMENTS FOR INSIDE CAVITIES

CASE #1 - MEASURING PARALLEL TO THE AXIS

$C = R \times \text{Cosec } B$
$E = \frac{1}{2} Y \times \text{Cotan } B$
$F = X - E$
$N = 1 + \text{Cosec } B$
$Y = 2 E \times \tan B$
$X = C + R$ ALSO:

$$\boxed{X = N \times R}$$

EXAMPLE : ∡ A = 60° ; D = ¾ IN.
SOLUTION : X = 3 × ⅜ = 1.125000 IN.

CASE #2 - MEASURING PARALLEL TO ONE SIDE

$G = R \times \text{Cotan } B$
$P = 1 + \text{Cotan } B$
$X = G + R$ OR:

$$\boxed{X = P \times R}$$

EXAMPLE : ∡ A = 75° ; D = 5⁄16 IN.
SOLUTION : X = 2.303210 × 5⁄32 = .359876 IN.

CONSTANTS FOR COMMON ANGLES

Angles		In Cases #1 & 6	In Cases #2 & 3	In Most Cases
A	B	N	P	T or cotan B
25°	12½°	5.620213	5.510714	4.510714
30°	15°	4.863708	4.732063	3.732063
35°	17½°	4.325574	4.171600	3.171600
40°	20°	3.923803	3.747466	2.747466
45°	22½°	3.613149	3.414288	2.414288
50°	25°	3.366203	3.144511	2.144511
55°	27½°	3.165679	2.920979	1.920979
60°	30°	3.000000	2.732051	1.732051
65°	32½°	2.861157	2.569701	1.569701
70°	35°	2.743375	2.428173	1.428173
75°	37½°	2.642667	2.303210	1.303210
80°	40°	2.555725	2.191747	1.191747

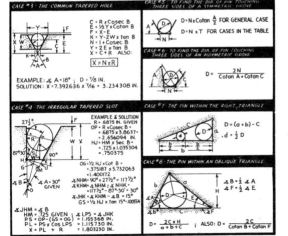

CASE #3 - THE COMMON TAPERED HOLE

$C = R \times \text{Cosec } B$
$E = \frac{1}{2} Y \times \text{Cotan } B$
$F = X - E$
$K = Y - 2W \times \tan B$
$N = 1 + \text{Cosec } B$
$Y = 2 E \times \tan B$
$X = C + R$ ALSO:

$$\boxed{X = N \times R}$$

EXAMPLE : ∡ A = 18° ; D = 7⁄8 IN.
SOLUTION : X = 7.392636 × 7⁄16 = 3.234308 IN.

CASE #4 - THE IRREGULAR TAPERED SLOT

EXAMPLE & SOLUTION
R = .6875 IN. GIVEN
OP = 2 × ∡ B =
 = .6875 × 3.8637 =
 = 2.656094 IN.
HJ = HM × Sec B =
 = .725 × 1.035304
 = .750375

OG = ½ HJ × Cot B =
 = .375187 × 3.732063
 = 1.400172
∡ NHM = 90° + 27½° = 117½°
∡ KHM = ∡ NHK =
 = 117½° − 87° = 30°
∡ JHK = ∡ KHM − ∡ B = 15°
GS = ½ HJ × tan 15° = .100554

∡ JHM = ∡ B
HM = .725 GIVEN ; ∡ LPS = ∡ JHK
PS = OP − (GS + OG) = 1.155368 IN.
PL = PS × cos LPS = 1.115730 IN.
X = PL + R = 1.803230 IN.

CASE #5 - TO FIND THE DIA. OF PIN TOUCHING THREE SIDES OF A SYMMETRIC GROVE

$D = N \times \text{Cotan } \frac{A}{2}$ FOR GENERAL CASE
$D = N \times T$ FOR CASES IN THE TABLE

CASE #6 - TO FIND THE DIA. OF PIN TOUCHING THREE SIDES OF AN ASYMMETRIC GROVE

$$D = \frac{2N}{\text{Cotan } A + \text{Cotan } C}$$

CASE #7 - THE PIN WITHIN THE RIGHT TRIANGLE

$D = (a + b) - c$
$d = \frac{1}{2} D$

CASE #8 - THE PIN WITHIN AN OBLIQUE TRIANGLE

∡ B = ½ ∡ A
∡ F = ¼ ∡ E

$$D = \frac{2C \times H}{a + b + c} \; ; \; \text{ALSO: } D = \frac{2C}{\text{Cotan } B + \text{Cotan } F}$$

> **Sec. VII**
> **Weights,**
> **Gages,**
> **Tolerances**

165

Gage Block Information

Precision gage blocks are manufactured to various qualities of precision. The finest quality is used as a laboratory set and is accurate to within .000002. This set is marked AA grade. The next quality is used as an inspection set and is accurate to within .000004. This set is marked A grade. The next quality is used as a working set and is accurate to within .000008. This set is marked B grade.

Typical Gage Block Series

First: .0001 Series -- 9 Blocks
.1001 .1002 .1003 .1004 .1005 .1006 .1007 .1008 .1009

Second: .001 Series -- 49 Blocks
.101 .102 .103 .104 .105 .106 .107 .108 .109 .110 .111
.112 .113 .114 .115 .116 .117 .118 .119 .120 .121 .122
.123 .124 .125 .126 .127 .128 .129 .130 .131 .132 .133
.134 .135 .136 .137 .138 .139 .140 .141 .142 .143 .144
.145 .146 .147 .148 .149

Third: .050 Series -- 19 Blocks
.050 .100 .150 .200 .250 .300 .350 .400 .450 .500 .550
.600 .650 .700 .750 .800 .850 .900 .950

Fourth: 1.000 Series -- 4 Blocks
1.000 2.000 3.000 4.000
Two .050 Wear Blocks

The method of selecting the proper gage blocks from a set to give a certain height dimension is as follows. From the number of blocks available first select a block from the set which will eliminate the digit in the fourth decimal place. Subtract the thickness of this block from the total height dimension. This is the first of several steps.

Next select a block to eliminate the digit in the third decimal place of the remainder. Subtract the thickness of this block from the previous remainder. Continue this process until there is no remainder. Always use as small a number of blocks as possible. On occasion two stacks of gage blocks are required to be the same height dimension. In this case a different set of blocks is selected because there is only one block for each size available in a set. Two blocks will be necessary to eliminate the fourth digit from the height dimension. See sample problem #2 stack.

Sample Problem: Select two stacks of gage blocks from a standard series set of gage blocks containing 83 blocks. See page 144 to give a height dimension of 2.0074.

Solution:

Stack #1		Stack #2	
2.0074		2.0074	
.1004	#1 block	.1003	#1 block
1.9070		1.9071	
.107	#2 block	.1001	#2 block
1.8000		1.8070	
.800	#3 block	.117	#3 block
1.0000		1.6900	
1.0000	#4 block	.140	#4 block
0.0000		1.5500	
		.950	#5 block
		.6000	
		.600	#6 block
		0.0000	

It is possible to obtain dimensions more accurate than tenths of a thousand, if a pair of tapered gage blocks, which are graduated with a vernier scale, are available. These blocks can be set to within .000010 by properly setting the blocks to the calibrated lines. It is possible to set the blocks to half or quarter graduations and get an estimated dimension of .000005 or less.

METRIC GAGE BLOCKS

Metric gage blocks are available and manufactured under precision standards the same as those required for inch dimensions. The procedure used for the selection of the blocks for a certain height dimension is also similar. The sizes of a typical metric gage block set consisting of 86 blocks is shown below.

A TYPICAL METRIC GAGE BLOCK SERIES

First:	.5 mm	1 block

Second:	1.001 mm through 1.009 mm	9 blocks

1.001, 1.002, 1.003, 1.004, 1.005, 1.006, 1.007, 1.008, 1.009.

Third:	1.01 mm through 1.49 mm	49 blocks

1.01, 1.02, 1.03, 1.04, 1.05, 1.06, 1.07, 1.08, 1.09, 1.10, 1.11, 1.12, 1.13, 1.14, 1.15, 1.16, 1.17, 1.18, 1.19, 1.20, 1.21, 1.22, 1.23, 1.24, 1.25, 1.26, 1.27, 1.28, 1.29, 1.30, 1.31, 1.32, 1.33, 1.34, 1.35, 1.36, 1.37, 1.38, 1.39, 1.40, 1.41, 1.42, 1.43, 1.44, 1.45, 1.46, 1.47, 1.48, 1.49.

Fourth:	1 mm through 9.5 mm	18 blocks

1, 1.5, 2, 2.5, 3, 3.5, 4, 4.5, 5, 5.5, 6, 6.5, 7, 7.5, 8, 8.5, 9, 9.5.

Fifth:	10 mm through 90 mm	9 blocks

10, 20, 30, 40, 50, 60, 70, 80, 90.

Note: See pages 166 and 167 for information as to the proper procedure for selecting blocks.

Example: Select two stacks of gage blocks from the set above for a height dimension of 46.034 mm.

Stack #1			Stack #2		
46.034			46.034		
− 1.004	#1 block		− 1.003	#1 block	
45.030			45.031		
− 1.030	#2 block		− 1.001	#2 block	
44.000			44.030		
− 4.000	#3 block		− 1.020	#3 block	
40.000			43.010		
− 40.000	#4 block		− 1.010	#4 block	
00.000			42.000		
			− 2.500	#5 block	
			39.500		
			− 9.500	#6 block	
			30.000		
			− 30.000	#7 block	
			00.000		

NOTES

Carbon Content of Carbon Steels for Different Uses

Carbon Range Per Cent	Uses of Carbon Steel
0.02–0.12	Chain, stampings, rivets, nails, wire, pipe; where very soft, plastic steel is needed.
0.10–0.20	Structural steels, machine parts. Soft and tough steel. For case-hardened machine parts, screws.
0.20–0.30	Gears, shafting, bars, bases, levers, etc. A better grade of machine and structural steel.
0.30–0.40	Lead screws, gears, worms, spindles, shafts, machine parts. Responds to heat treatment, but is often used in the natural condition.
0.40–0.50	Crankshafts, gears, axels, mandrels, tool shanks, and heat treated machine parts.
0.60–0.70	Low carbon tool steel, used where a keen edge is not necessary, but where shock strength is wanted. Drop hammer dies, set screws, screw drivers, arbors.
0.70–0.80	Tough and hard steel. Anvil faces, band saws, hammers, wrenches, cable wire, etc.
0.80–0.90	Punches for metal, rock drills, shear blades, cold chisels, rivet sets, and many hand tools.
0.90–1.00	Used for hardness and high tensile strength, springs, cutting tools, press tools, and striking dies.
1.00–1.10	Drills, taps, milling cutters, knives, etc.
1.10–1.20	Drills, taps, knives, cold cutting dies, wood working tools.
1.20–1.30	Files, reamers, knives, tools for cutting wood and brass.
1.30–1.40	Used where a keen cutting edge is necessary; razors, saws, instruments and machine parts where maximum resistance to wear is needed. Boring and finishing tools.

Hardening of Carbon Tool Steel

Carbon tool steel may be hardened by heating slowly to a temperature range extending from 1400°F. to 1550°F., followed by quenching in cold water or brine. It is recommended that a brine solution made by dissolving approximately one pound of salt (sodium chloride, or rock salt) to a gallon of water be used for the quenching bath. In order to obtain a uniform quenching action, the quenching bath should be agitated, or at least the piece to be quenched should be moved around in the quenching bath during the hardening process. In event the piece to be quenched is placed in the quenching medium without any agitation, soft spots may be the result.

A careful study of the manner of heating, the temperature to be used, and the time at temperature should be made before attempting to harden a piece of tool steel.

A furnace should be selected that will heat the steel to the proper temperature without causing too much scale or decarburization. Oven furnaces heated with gas or electricity may be successfully used with the proper care. Also, salt or lead baths serve as good heating mediums. It is advisable to have any furnace equipped with good temperature indicators or recorders, and it is very helpful to have the furnace operate with automatic temperature controls. The automatic controls will greatly facilitate the operation of the furnace and insure more accurate control of the temperature.

In general, most of the common carbon tool steels may be successfully hardened from a temperature of 1450°F., but the hardening temperature may be varied depending upon the type of steel, and upon the size of tool to be hardened. Also, the use to which the tool is to be put influences the hardening temperature. A small tool made from a standard grade of carbon steel may be hardened successfully from 1400°F. in brine, or even by quenching in oil. However, if oil is to be used it is advisable to use a higher temperature. Perhaps a temperature of 1500°F. may be found satisfactory. It will be observed that the higher temperature will develop a coarser grain, but will cause a deeper hardening effect and easier hardening. It should be borne in mind that coarsening of the grain will result in less toughness in the hardened steel and more danger from cracking and warping. A good rule to follow is never heat the steel to a higher temperature than is necessary to develop the proper hardening effect.

Good instructions for the proper hardening methods to be used with any steel may be obtained from the steel manufacturer or supplier. The best practice is to read carefully the instructions received from this source and follow out the directions as to heating method, rate of heating, maximum temperature, time at heat, and the manner of quenching or cooling.

A successfully hardened tool made from a standard grade of carbon tool steel will test approximately 65 Rockwell C on the standard Rockwell Hardness Tester, or if tested with a file should be file hard. (File hard means that the steel cannot be scratched or filed with a standard file.)

Critical Temperature Diagram

The following diagram illustrates the temperature ranges that may be selected for the heat treating of carbon steels

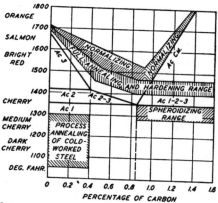

NOTES: 1. HEATING SHOULD BE SLOW AND UNIFORM
2. FOR BEST RESULTS NEVER HEAT HIGHER THAN NECESSARY
3. NORMALIZING = AIR COOLING
4. ANNEALING = FURNACE OR PACK COOLING
5. HARDENING = QUICK COOLING

such as hardening, annealing, normalizing, etc., and is called the "critical temperature diagram" because it shows the temperatures where the changes take place upon heating the various carbon steels, which allows us to carry out the various heat treating operations.

TEMPERING OR DRAWING OF
HARDENED STEEL

Fully hardened tool steel with a carbon content greater than 0.60% carbon is brittle and therefore dangerous as a tool or structural part of a machine. Any slight overload or sudden shock load may cause failure. The heat treater, by means of tempering or drawing, may reduce the brittleness and increase the toughness, therefore lessening the danger of breakage. Tempering of steel consists of heating the quenched-hardened steel to some pre-determined temperature between room temperature and about 1000° F. for a short time, followed by air cooling. The time at the tempering temperature may vary from a few minutes to about 2 hours. One hour tempering time is usually recommended. The following chart may be followed to indicate the approximate temperatures to use in the tempering operation:

Tempering or Drawing Chart

Degrees Fahrenheit	Suggested Uses for Carbon Steels
350–400	Tools for metal cutting that must be of maximum hardness, drills, taps, paper knives, lathe tools, etc.
400–450	Tools that need hardness and more toughness, rolled thread dies, punches and dies.
450–500	Rock drills, hammer faces, shear blades, and tools where toughness is required.
550–600	Axes, knives, iron and steel chisels, saws for wood, and in tools that may be sharpened and shaped by use of a file.
650–700	This temperature is usually too high for cutting tools and dies, but may be used for tempering springs.

Tempering of steel may be accomplished in liquid baths such as oil, salt, lead, or in suitably designed oven furnaces. If a furnace is not available, tempering may be carried out by the use of so-called "temper colors." The hardened steel is polished and then heated, the brightened surface will take on colors due to formation of thin films of oxide. These color oxides may be used as a rough indication of temperatures, and serve as a means of measuring the temperature to be used in the tempering operation. The following table indicates the approximate color oxide formed at the corresponding tempering temperature:

173

Degrees F.	Oxide Color
430	Faint Straw
460	Dark Straw
500	Bronze
540	Purple
570	Dark Blue
610	Light Blue
630	Steel Gray

In using the temper color method for tempering, the hardened and polished steel is slowly heated over a fire or any hot medium until the color corresponding to the desired temperature is seen on the polished surface. Care should be taken to prevent heating beyond the temperature desired. Cooling may be carried out in oil or in air. This method should never be recommended in preference to the use of a tempering furnace.

Warping and Cracking

Warping and cracking are serious menaces in nearly every hardening and tempering operation. Warping or cracking is caused by the severe stresses set up by the uneven contraction and expansion that takes place during the hardening operation. Distortion or warping will always take place; and if the stresses become severe enough, breakage by cracking may occur. A study of the part to be hardened may avoid a lot of costly straightening or grinding, or perhaps remove the danger of cracking. The heat treater may exhibit real skill and craftsmanship in the manner he selects to treat a given steel object to be hardened. Some of the factors that contribute to warping and cracking may include: (1) Unbalanced and abrupt changes in section; (2) Sharp corners and deep tool marks, which act as crack formers; (3) Intricate shapes, with many cut out sections; (4) Overheating the steel in hardening; (5) Defects in the steel, such as seams, laps, cracks, etc.

Soft Spots

One of the most common troubles resulting from the hardening of steel is a soft skin on the surface of the hardened tools, or soft spots on the faces of the steel. Soft skin is the result of not removing enough stock from the surfaces of the steel as received from the steel mill, or may be due to decarburization occurring during the hardening operation. Any decarburization will cause a soft skin. Soft spots are the result of poor quenching technique, preventing the steel from cooling fast enough in the quenching medium for proper hardening. Violent agitation in a brine quenching bath will usually remove the trouble and prevent the formation of soft spots.

Oil and Air Hardening Steels

The so-called carbon steels, in general, are hardened by quenching in water or brine. However, if the section to be hardened is thin—less than $\frac{1}{4}$ inch—carbon steels may be often hardened by quenching in oil and develop full hardness. In event the section is such as to allow oil quenching, it is highly recommended. The oil hardened steel will be less distorted and less likely to crack. If the section is such as to require water hardening of a carbon steel and oil quenching is desired, a special steel or alloy steel may be selected that will harden fully in an oil quench. Such steels may be hardened successfully in oil or even air. It is recommended that in event an alloyed oil-or-air-hardening steel is to be tried, the user consult with his steel supplier and follow his recommendations as to the type of steel to buy and as to the method of heat treating this special steel. Some of the reasons for the selection of a special or alloy steel in preference to a carbon steel may include the following:

1. Because of oil or air hardening characteristics, alloy steels are less apt to crack and are in less danger from warping.

2. Alloy steels are deeper hardening, and hardening heavy sections is apt to be more satisfactory with them than with a carbon steel.

3. Alloy steels are more stable upon heating and resist softening which allows the use of hardened alloy steels in dies and tools subjected to heating.

4. Some alloy steels develop high resistance to shock loading and are tougher than carbon steels of the same hardness.

Hardness of Steel

The hardness of steel is largely controlled by the carbon content of the steel, and the heat treatment it has received. In other words, the hardness of any steel in the fully hardened condition will depend chiefly upon its carbon content, regardless of the amount of special alloy element present in the steel. As measured by the standard hardness testing machines, a Rockwell Hardness of 67 C is about the maximum hardness any plain carbon or alloy steel develops. Hardness may be measured by several different methods, and in the shop where a hardness testing machine is available an accurate gage of the results of any heat treatment may be checked by such a machine. If a hardness testing machine is not available, testing for maximum hardness may be carried out by the use of a file. When a steel cannot be filed it is pronounced "file hard" and considered to be in the hardest possible state. There is no definite relationship between the hardness as measured by the different methods. However, tests have been made and various conversion tables have been constructed which are useful when approximate conversion values are needed.

Quenching Baths for Hardening

Steel may be quenched in the following baths for the purpose of hardening:

1. Hot or Molten Salt Bath
2. Air Blast
3. Emulsion of 10% Oil in Water
4. Water at 122°F.
5. Oil
6. Water at 65°F.
7. Brine—Aqueous Solution of 5% NaCl
8. Caustic Bath—Aqueous Solution of 3% NaOH

These baths cool at a faster rate in the order given. Violent agitation always increases the rate of quenching of any media.

Normalizing of Steel

Steel that is in a poor condition such as having a coarse grain size, or having a non-uniform and banded structure found often in forgings, may be given a normalizing heat treatment to improve it. The normalizing consists of heating the steel into the normalizing temperature range shown in the **Critical Temperature Diagram,** followed by air cooling. This treatment should result in an improvement in the uniformity resulting from any further heat treatments such as hardening.

Annealing of Steel

Tool steel may be fully annealed by heating to the annealing temperature range, 1400–1475°F., followed by very slow cooling. The slow cooling may be carried out by allowing the steel to cool with the furnace, or by packing the steel in boxes or pots and allowing the steel to cool from the annealing temperatures in the box. The slower the cooling, the softer will be the steel. To avoid scaling or decarburization, the steel should be packed in a suitable carbonaceous material so as to prevent excessive oxidation. Cast iron chips make a good packing material. If a furnace is available that has a controlled atmosphere, annealing may be carried out without the use of a packing material; and the anneal will be without scaling—in fact maintain a bright surface on the steel. Furnaces for bright annealing are of the muffle type and employ an atmosphere that is very reducing and rich in carbon monoxide, nitrogen or hydrogen.

Case Hardening

Probably the oldest of heat treatments is that of case hardening, i.e., taking a low carbon and soft steel, and heating it to about 1650°F. in a carbonaceous atmosphere followed by quenching in water or brine. This treatment results in the formation of a high carbon layer, called the "case" on the low carbon steel; and upon quenching, the high carbon "case" becomes file hard. Various carburizers are available for this treatment, the base for which is usually charcoal. The treatment consists of packing the steel in a box with the carburizing compound, and heating the box in a furnace to about 1650°F. for a predetermined length of time. The longer the steel is held at the carburizing temperature, the deeper will be the penetration of the carbon (case). In general, we may expect to obtain a penetration of approximately 0.006 inch per hour of treatment. After the carburizing treatment the steel may be quenched directly from the carburizing heat, or may be cooled slowly to room temperature followed by reheating to a temperature of 1400–1450°F. and quenched. Less distortion and a finer grained case results from the reheating method. This case hardening treatment is used to reduce the cost of making steel parts that need surface hardness, and where the part requires a combination of hard surface and tough core for greater shock strength than can be obtained from the hardening of a tool steel.

Case hardening may be carried out in a gaseous atmosphere rich in carbon monoxide, or in a nitriding atmosphere as in the nitriding process. Also, case hardening may be carried out through the use of liquid carburizers, such as the cyanide salt bath.

Many small parts, made from low carbon steel, that require a high surface hardness with only a light or thin case, may be successfully carburized and hardened in the cyanide liquid baths. Steel to be case hardened is immersed in a molten bath containing more than 25% sodium cyanide held at a temperature usually around 1550°F. At this temperature, the steel will pick up both carbon and nitrogen from the bath and in 15 minutes a penetration of approximately 0.005 inch is obtained. If deeper penetration is required, a longer time is necessary. A treatment of one hour results in about 0.010 inch depth of case. To harden the case, quenching direct from the carburizing temperature in water or brine is practiced. If less distortion is needed than from the usual practice, lower carburizing temperatures may be employed or cooling to a lower temperature before quenching.

All parts placed in the molten bath must be free from moisture or danger from explosive spattering of the bath may occur. The operator should be well protected by gloves and goggles or helmet. The fumes given off by the bath, because of their obnoxious nature, should be removed by means of a hood over the bath and a suitable exhaust system.

Flame Hardening

Flame hardening is used to surface-harden large and small steel and gray cast-iron machine parts, parts that would not permit heating in a furnace and water or oil hardening, due to danger of warping and cracking. In this process of surface hardening, heat is applied to the surface of the steel or cast-iron by an oxy-acetylene flame. Only a thin layer of the surface of the metal is brought up to the hardening heat, and as the torch moves slowly forward heating the metal, a stream of water follows the torch, quenching and hardening the surface as it becomes heated. The speed of the torch is adjusted so that the heat only penetrates to the desired depth, thereby controlling the depth of hardness. Small parts may be individually heated and quenched. In the case of cylindrical shapes such as shafts, the surface may be heated by slow rotation of the shaft and exposing the surface to the flame of a torch. Upon reaching the hardening temperature and when the heat has penetrated to the desired depth, the shaft may be quenched for hardening. In general, a steel to be hardened by this method should have a carbon content of at least 0.30%. The best range of carbon is between 0.40% and 0.70% carbon. Higher carbon steels may be treated by flame hardening, but care must be exercised to prevent surface cracking or checking. Applications of the flame-hardening process include rail ends, gears, lathe beds, track wheels for conveyors, cams and cam shafts, jaw clutches on large machines, etc.

Induction Hardening

In the induction hardening process the surfaces of the part to be hardened are heated by means of inductor heating coils placed around the surfaces. The inductor coils induce a current in the surface of the steel. This induced current rapidly heats the metal to the proper hardening temperature. When the area to be hardened has been subjected to an accurately controlled current of high frequency for the correct time, the electric circuit is opened, and simultaneously the heated surface is quenched by a spray of water from a water jacket attached to the fixture holding the heating coils.

The time cycle is only a few seconds for the complete operation of heating and quenching. This process may be applied to many parts of machines and tools, provided the carbon content is high enough to permit quench hardening. This process is used in the hardening of many parts of machine tools such as gears, cams, shafts, clutches, crankshafts, etc. Its great advantage lies in the freedom from distortion that may be had over the older methods used to case harden and heat treat steel parts. Gears may be hardened by this process without grinding after hardening to produce accurate dimensions.

HIGH-SPEED STEELS

These steels are so named because when used as tools, such as lathe tool bits, they can remove metal much faster than ordinary steel. Their chief superiority over other steels lies in their ability to maintain their hardness even at a dull red heat (about 1100°F.). Due to this property, such tools may operate satisfactorily at speeds which cause the cutting edges to reach a red heat. These steels contain up to 20% tungsten, from 2% to 5% chromium, usually from 1% to 2% vanadium, and sometimes several per cent of cobalt. More recently, molybdenum has been successfully substituted for the tungsten in these steels, with molybdenum up to 9% and no tungsten, although a more common high-speed steel contains about equal amounts of molybdenum and tungsten (5% tungsten and 5% molybdenum). The high-speed steel that has been most commonly used is referred to as 18–4–1 which means that it contains 18% tungsten, 4% chromium, and 1% vanadium. The carbon content in the usual high-speed steel runs about 0.65% to 0.75%.

The molybdenum or molybdenum-tungsten high-speed steels have been largely substituted for the straight tungsten high-speed steels of the 18–4–1 type; and actual shop tests made with heavy and light lathe tools, planer tools, milling reamers demonstrated that light and particularly heavy planer and lathe tools stood up, on the average, as well or better than tungsten or the tungsten-cobalt tools. The molybdenum grades of high-speed steels have one drawback to their use in that they are much more sensitive to decarburization than the tungsten steels during the hot-working and heat-treating operations. This problem can be overcome by the use of borax, which is fused in a thin coating over the steel, forming a protective film during the heating for hot forging and during subsequent forging operations. The borax may also be used during the heat treating operations, however it causes a cleaning problem, and is injurious to the furnace hearth. In the place of borax, a furnace designed with the correct atmosphere will prevent any serious decarburization. Also, it has been found that it is practical to use salt baths for the heat-treating operations, the molten salt protecting the surfaces of the molybdenum high-speed steel from decarburization.

Heat Treating of High-speed Steels

Heating for hardening should be carried out in controlled atmosphere furnaces, or in salt baths, or the tools before heating may be dipped in a saturated solution of borax in water heated to between 150°F. and 180°F. Tools dipped in this warm solution will dry when removed, and be covered with a thin powdery film of borax that will not rub off if the tools are handled with moderate care. If too much borax is used, blister may form on the surface of the hardened tools. After hardening, the tools will be found coated with a thin layer of borax which may be removed with a wire brush, sand blast, or 10% solution of acetic acid.

Preheating for Hardening

It is recommended that high-speed steels be preheated to a temperature range from 1400°F. to 1500°F. before placing the tools in the hardening furnace. This preheating should be carried out in a furnace separate from the one used for the hardening heat. The molybdenum high-speed steels may be preheated to a lower temperature than the straight tungsten high-speed steels which seems to help in avoiding decarburization. Small tools of light section, under ¼ inch, may be hardened without preheating, but all heavy sections should be fully preheated.

Hardening Temperatures for High-speed Steels

The straight tungsten type of high-speed steel, such as 18–4–1, may be hardened by heating to a temperature range from 2300°F. to 2375°F., followed by quenching in an oil bath. The time required to heat will depend upon the size of section and upon the type of heating medium. The timing should be such as to allow the steel to come up to the temperature of the furnace, or nearly so. For light sections, this may take only a few minutes; whereas, for a section of 1 to 2 inches it may require a heating time of 20 minutes. Careful observation of the steel during the heating cycle will allow the hardener to judge the correct time required for heating. Anyone who has hardened the 18–4–1 type of tungsten high-speed steel can harden molybdenum high-speed steels. The hardening is carried out in a similar way, preheating at 1200°F. to 1500°F. and transferring the pre-heated steel to the high temperature furnace for hardening. The quenching and drawing are done in a similar way. The steel should never be soaked excessively at the hardening temperature, but should be removed from the furnace and quenched very shortly after it is up to heat. The temperatures recommended for the hardening of the molybdenum type of high-speed steel are somewhat lower than for the tungsten steels, i.e., 2150°F. to 2250°F. Overheating of the molybdenum steels will result in a somewhat brittle condition, whereas 18–4–1 may be overheated 75°F. and still make a reasonably satisfactory tool.

Quenching of High-speed Steels

The quenching may be done in oil, salt bath, or air for both the tungsten and molybdenum type steels. Oil is the most common quenching medium. However, if warping or cracking becomes a problem it is recommended that hot quenching be used, quenching in molten salts held at a temperature range from 1100°F. to 1200°F., or quenching in air if scaling is not serious and the section hardens satisfactorily. A practice that may be tried is to quench in oil for a short time then placing the tool in the preheat furnace held at 1200°F. until the temeperature becomes equalized, followed by air cooling to room temperature. The time in the oil quench should be such as to cool the steel from the high heat to around 1200°F., or a very dull red heat.

Tempering or Drawing of High-speed Steels

High-speed steels after hardening do not develop their full hardness and toughness without tempering or drawing. Drawing may be successfully carried out by heating the hardened steel to a temperature range extending from 1025°F. to 1100°F. for at least 1 hour, preferably 2 hours. The steel before tempering should be cooled to at least 200°F. from the quench, and reheated slowly to the temperature used in drawing. The cooling from the drawing temperature may be done in air. **Do not quench in water.** Multiple tempering is recommended when maximum toughness is required, i.e., repeat the first tempering operation.

Hardness of High-speed Steels

High-speed steels in the fully annealed condition will Rockwell around C–10 to C–20, and are in a machinable condition. In the hardened condition, as quenched, the hardness should run between Rockwell C–63 and C–65 for satisfactory results. After drawing, the hardness should run between Rockwell C–64 and C–66.

Annealing of High-speed Steels

Annealing may be carried out by slowly heating the high-speed steel to a temperature from 1550°F. to 1675°F. for about 6 to 8 hours, followed by very slow cooling of the steel. The cooling cycle should require about 24 hours. The steel must be protected from oxidation and decarburization by packing in cast iron chips or a suitable reducing agent such as charcoal and ashes mixed. As most high-speed steel is purchased in the annealed condition for machining, it is not necessary to anneal this type of steel unless reannealing and rehardening of a used tool is required. However, if this is the case, the steel should be reannealed before hardening, otherwise a brittle condition of the rehardened tool will result.

IRON AND STEEL DEFINITIONS

Pig Iron. Pig iron is the product of the blast furnace and is made by the reduction of iron ore.

White Cast Iron. White cast iron is cast iron with the carbon in the combined form, which makes the iron hard and brittle, and fractures with a white fracture appearance.

Gray Cast Iron. Gray cast irons are cast irons with some combined carbon and the balance as graphite. The term "gray iron" is derived from the characteristic gray fracture of this metal.

Malleable Cast Iron. This term is applied to castings in which all the carbon in a special white cast iron has been changed to free or temper carbon (graphite) by suitable annealing treatment.

Alloy Cast Iron. This term applies to cast irons which have been alloyed with such elements as nickel, chromium, molybdenum, etc.

Wrought Iron. Wrought iron is an iron of high purity containing a particular type of glass-like slag (iron silicate).

Steel. Steel is a cast metal containing from 0.04% to 1.50% carbon that is malleable and workable as cast.

Acid Steel. Steel melted in a furnace with an acid (siliceous) bottom and lining and under a slag that is dominantly siliceous.

Basic Steel. Steel melted in a furnace with a basic bottom and lining and under a slag which is dominantly basic.

Killed Steel. A steel sufficiently deoxidized to prevent gas evolution during solidification.

Rimmed Steel. An incompletely deoxidized steel normally containing less than 0.25% carbon which on solidification liberates gases keeping the top of the steel open until the bottom and sides solidify.

Carbon Steel. Steel which owes its properties chiefly to the various percentages of carbon without substantial amounts of other alloying elements.

Cast Steel. Any object made by pouring molten steel into molds.

Alloy Steel. Steel which owes its distinctive properties chiefly to some element or elements other than carbon or jointly to such other element and carbon.

Oil Hardening Steel. An alloy steel that will develop its maximum hardness upon quenching in oil from the hardening temperature.

Air Hardening Steel. An alloy steel which may be quenched in still air from the hardening temperature and will develop its full hardness.

Hardening. Heating and quenching of steel from above its critical temperature for purpose of producing a hardness superior to that when the alloy is not quenched.

Tempering (also termed drawing). Reheating hardened steel to some temperature below the critical temperature, followed by any desired rate of cooling.

Annealing. A heating and cooling operation implying a relatively slow rate of cooling.

Process Annealing. Heating steel to a temperature usually below the critical temperature range followed by cooling as desired.

Normalizing. Heating steel to about 100°F. above the critical temperature range followed by cooling in air.

Patenting. Heating steel to above the critical range followed by cooling in molten lead or molten salts held at 800°F. to 1050°F., or by cooling in air.

Spheroidizing. Any process of heating and cooling steel that produces a rounded or globular form of carbide.

Austenite. Solid solution of iron and carbon in steel in which gamma iron is the solvent.

Martensite. The structure of fully hardened steel is known as "martensite." It is of maximum hardness and has a needle-like structural pattern.

Troostite. A fine aggregate structure of ferrite and cementite that is not resolved under the microscope. It is softer than martensite.

Sorbite. Sorbite is a structure formed by tempering martensite to a relatively high temperature. The structure is distinctly granular.

Pearlite. A laminated structure of ferrite and carbide resulting from the direct transformation of austenite at the lower critical temperature upon cooling.

Cementite. A carbide of iron having the formula Fe_3C. The hardener in steel.

Ferrite. The term "ferrite" is applied to solid solutions in which alpha iron is the solvent.

Alpha Iron. Alpha iron is the body-centered cubic crystal state of iron and is strongly magnetic.

Gamma Iron. A crystal form of iron (face-centered cubic) that forms when iron or steel is heated above its critical temperature. Gamma iron is non-magnetic.

183

GRINDABILITY

EASY TO GRIND . . . . . . HARDER TO GRIND

		HIGH GRINDABILITY	MEDIUM GRINDABILITY	LOW GRINDABILITY	HIGH VANADIUM
TOOL STEEL GROUPS		S Shock Resisting H Hot Work O Oil Hardening L Low Alloy	A Air Hardening W Water Hardening D High Carbon- High Chrome (Exc. Hi-Vanadium)	T High Speed Steel M High Speed Steel (Exc. Hi-Vanadium)	D7 - M4 - M15 T9 - T15 4% or more Vanadium
	GRINDING WHEEL SPECIFICATION				
	Grain Type	Semi-Friable Aluminum Oxide	Friable Aluminum Oxide	Friable Aluminum Oxide	Friable Aluminum Oxide
	Grit Size Grade Hardness Bond Type	36-60 H-J Vitrified	36-60 G-H Vitrified	46-60 F-G Vitrified	80-100 F Vitrified
	SPEEDS AND FEEDS Wheel Speed Table Speed Crossfeed Infeed	5500-6500 SFPM 50-75 FPM 1/16"-1/32" .001"-.005"	5500-6500 SFPM 50-75 FPM 1/16"-1/2" .001"-.003"	5500-6500 SFPM Fast as possible 1/32"-1/64" .0005"-.001"	3000 SFPM Fast as possible Full face of wheel .0005"-.001"

SURFACE GRINDING

Sec. VIII
General
Information

184

GRINDING WHEEL SPECIFICATION		Semi-Friable Aluminum Oxide	Friable Aluminum Oxide	Friable Aluminum Oxide	Friable Aluminum Oxide or Green Silicon Carbide
CYLINDRICAL GRINDING	Grain Type	Semi-Friable Aluminum Oxide	Friable Aluminum Oxide	Friable Aluminum Oxide	Friable Aluminum Oxide or Green Silicon Carbide
	Grit Size	60-80	60-80	60-80	80-100
	Grade Hardness	K-L	J-K	H-J	F-H
	Bond Type	Vitrified	Vitrified	Vitrified	Vitrified
	SPEEDS AND FEEDS				
	Wheel Speed	5500-6500 SFPM	5500-6500 SFPM	5500-6500 SFPM	3000 SFPM
	Work Speed Roughing	50-90 SFPM	50-90 SFPM	70-100 SFPM	70-100 SFPM
	Finishing	90-150 SFPM	90-150 SFPM	100-150 SFPM	100-150 SFPM
	Infeed	.001"-.002"	.001"-.002"	.001"-.002"	.0005"-.001"
	Table Speed Roughing	1/2-1/3 width of wheel per Rev. of work	1/2-1/3 width of wheel per Rev. of work	1/3-1/6 width of wheel per Rev. of work	1/3-1/6 width of wheel per Rev. of work
	Finishing	1/3-1/6 width of wheel per Rev. of work	1/3-1/6 width of wheel per Rev. of work	1/6-1/12 width of wheel per Rev. of work	1/6-1/12 width of wheel per Rev. of work
GRINDING FLUID	Straight Grinding	Water Miscible Fluid	Water Miscible Fluid	Water Miscible Fluid	Water Miscible Fluid
	Complex Form Grinding	Straight Oil	Straight Oil	Straight Oil	Straight Oil

185

STANDARD MARKING SYSTEM CHART

FOR IDENTIFYING GRINDING WHEELS

Sequence	1	2	3	4	5	6
	Prefix	Abrasive Type	Grain Size	Grade	Structure	Bond Type	Manufacturer's Record
	51	A	36	L	5	V	23

Prefix: MANUFACTURER'S SYMBOL INDICATING EXACT KIND OF ABRASIVE. (USE OPTIONAL)

Abrasive Type:
ALUMINUM OXIDE—A
SILICON CARBIDE—C

Grain Size:

Coarse	Medium	Fine	Very Fine
10	30	70	220
12	36	80	240
14	46	90	280
16	54	100	320
20	60	120	400
24		150	500
		180	600

Grade: GRADE SCALE

Soft
A B C D E F G H I J K L M N O P Q R S T U V W X Y Z
Hard

Medium (at L–M region)

Structure:

Dense to Open	
1	9
2	10
3	11
4	12
5	13
6	14
7	15
8	Etc.

(USE OPTIONAL)

Bond Type:
V—VITRIFIED
S—SILICATE
R—RUBBER
RF—RUBBER REINFORCED
B—RESINOID
BF—RESINOID REINFORCED
E—SHELLAC
O—OXYCHLORIDE

Manufacturer's Record: MANUFACTURER'S PRIVATE MARKING TO IDENTIFY WHEEL. (USE OPTIONAL)

HARDNESS CONVERSION TABLE—APPROXIMATE

BRINELL		ROCKWELL				
Diam. of Impression For 3000 Kg. Load and 10 mm Ball	Hardness No.	C Scale 150 Kg. 120° Diamond Cone	B Scale 100 Kgs. ☆ Ball	SHORE	Vickers or Firth Diamond Hardness Number	Tensile Strength— 1000 lbs. sq. inch.
2.20 mm	780	68		96	1220	
2.25	745	67		94	1114	
2.30	712	65		92	1021	354
2.35	682	63		89	940	341
2.40	653	62		86	867	329
2.45	627	60		84	803	317
2.50	601	58		81	746	305
2.55	578	56		78	694	295
2.60	555	55		75	649	284
2.65	534	53		73	608	273
2.70 mm	514	51		71	587	263
2.75	495	50		68	551	253
2.80	477	48		66	534	242
2.85	461	47		64	502	233
2.90	444	46		62	474	221
2.95	429	44		60	460	211
3.00	415	43		58	435	202
3.05	401	42		56	423	193
3.10	388	41		54	401	185
3.15	375	39		52	390	178
3.20 mm	363	38		51	380	171
3.25	352	37		49	361	165
3.30	341	36		48	344	159
3.35	331	35		46	335	154
3.40	321	34		45	320	148
3.45	311	32		43	312	143
3.50	302	31		42	305	139
3.55	293	30		41	291	135
3.60	285	29		40	285	131
3.65	277	28		38	278	127
3.70 mm	269	27		37	272	124
3.75	262	26		36	261	121
3.80	255	25		35	255	117
3.85	248	24	100	34	250	115
3.90	241	23	99	33	240	112
3.95	235	22	99	32	235	109
4.00	229	21	98	32	226	107
4.05	223	20	97	31	221	105
4.10	217	18	96	30	217	103
4.15	212	17	95	30	213	100
4.20 mm	207	16	95	29	209	98
4.30	197	14	93	28	197	95
4.40	187	12	91	27	186	91
4.50	179	10	89	25	177	87
4.60	170	8	87	24	171	84
4.70	163	6	85	23	162	81
4.80	156	4	83	23	154	78
4.90	149	2	81	22	149	76
5.00	143	0	79	21	144	74
5.10	137	−3	77	20	136	71

Sec. VIII
General
Information

High Temperatures Judged by Color, and Colors for Tempering

Degrees Centigrade	Degrees Fahrenheit	High Temperatures Judged by Color	Degrees Centigrade	Degrees Fahrenheit	Colors for Tempering
400	752	Red heat, visible in the dark	221.1	430	Very pale yellow
474	885	Red heat, visible in twilight	226.7	440	Light yellow
525	975	Red heat, visible in daylight	232.2	450	Pale straw-yellow
581	1077	Red heat, visible in sunlight	237.8	460	Straw-yellow
700	1292	Dark red	243.3	470	Deep straw-yellow
800	1472	Dull cherry-red	248.9	480	Dark yellow
900	1652	Cherry-red	254.4	490	Yellow-brown
1000	1832	Bright cherry-red	260.0	500	Brown-yellow
1100	2012	Orange-red	265.6	510	Spotted red-brown
1200	2192	Orange-yellow	271.1	520	Brown-purple
1300	2372	Yellow-white	276.7	530	Light purple
1400	2652	White welding heat	282.2	540	Full purple
1500	2732	Brilliant white	287.8	550	Dark purple
1600	2912	Dazzling white (bluish-white)	293.3	560	Full blue
			298.9	570	Dark blue

MACHINE-SHOP FORMULAS
(Miscellaneous)

1. To figure the amount to offset the tailstock for turning a taper on the lathe:

   O = offset; $T"$ = taper per inch; $L"$ = length of the work in inches

   $$\text{Formula} \qquad O = \frac{T" \times L"}{2}$$

2. To figure the angle to set the compound rest or the vertical head of a boring mill when the taper per inch is known:

   $T"$ = taper per inch

   $$\text{Formula} \qquad \frac{T"}{2} = \text{tangent of the angle}$$

3. To figure the change gears for helical milling:

   $L_1"$ = lead of helix to be cut; $L_2"$ = lead of the milling machine;
   d_t = product of the number of teeth in the driven gears;
   D_t = product of the number of teeth in the driving gears.

   $$\text{Formula} \qquad \frac{L_1"}{L_2"} = \frac{d_t}{D_t}$$

   Note: The following change gears are usually furnished with the Brown and Sharpe, Cincinnati, Milwaukee, and Ohio milling machines that are equipped for helical milling---two 24t., 28t., 32t., 40t., 44t., 48t., 56t., 64t., 72t., 86t., 100t.
   The Kempsmith milling machine is equipped with the above change gears except for an additional 70t. and 96t. instead of 100t.

4. When measuring standard threads with a thread micrometer:

   (a) The thread micrometer reading is equal to the pitch diameter of the thread less the tolerance desired.

   (b) The major diameter of the thread minus the single depth of the thread equals the pitch diameter of the thread.

   (c) To obtain the thread micrometer reading, subtract the single depth of the thread minus the tolerance desired from the major diameter.

Metric Measures

In accordance with the standard practice approved by the American Standards Association, the ratio 25.4 mm ÷ 1 inch is used for converting millimeters to inches. This factor varies only two millionths of an inch from the more exact factor 25.40005 mm, a difference so small as to be negligible for industrial length measurements.

The metric unit of length is the meter = 39.37 inches.

The metric unit of weight is the gram = 15.432 grains.

The following prefixes are used for sub-divisions and multiples:
Milli = $\frac{1}{1000}$, Centi = $\frac{1}{100}$, Deci = $\frac{1}{10}$, Deca = 10, Hecto = 100, Kilo = 1000, Myria = 10,000.

Metric and English Equivalent Measures

MEASURES OF LENGTH

Metric	English
1 meter	= 39.37 inches, or 3.28083 feet, or 1.09361 yards
.3048 meter	= 1 foot
1 centimeter	= .3937 inch
2.54 centimeters	= 1 inch
1 millimeter	= .03937 inch, or nearly 1-25 inch
25.4 millimeters	= 1 inch
1 kilometer	= 1093.61 yards, or 0.62137 mile

MEASURES OF WEIGHT

Metric	English
1 gram	= 15,432 grains
.0648 gram	= 1 grain
28.35 grams	= 1 ounce avoirdupois
1 kilogram	= 2.2046 pounds
.4536 kilogram	= 1 pound
1 metric ton } 1000 kilograms }	= { .9842 ton of 2240 pounds { 19.68 cwt. { 2204.6 pounds
1.016 metric tons } 1016 kilograms }	= 1 ton of 2240 pounds

MEASURES OF CAPACITY

Metric	English
1 liter (= 1 cubic decimeter)	= { 61.023 cubic inches { .03531 cubic foot { .2642 gal. (American) { 2.202 lbs. of water at 62° F.
28,317 liters	= 1 cubic foot
3.785 liters	= 1 gallon (American)
4.543 liters	= 1 gallon (Imperial)

METRIC MEASURE

10^3	10^2	10^1	10^0	10^{-1}	10^{-2}	10^{-3}
1000	100	10	1	$\dfrac{1}{10}$	$\dfrac{1}{100}$	$\dfrac{1}{1000}$
Thousands	Hundreds	Tens	Units	Tenths	Hundredths	Thousandths

meter

10^{-3}	meter	is a **milli**meter	(mm)
10^{-2}	meter	is a **centi**meter	(cm)
10^{-1}	meter	is a **deci**meter	(dm)
10^0	meter	is a **meter**	(m)
10^1	meters	is a **deka**meter	(dam)
10^2	meters	is a **hecto**meter	(hm)
10^3	meters	is a **kilo**meter	(km)

Prefix	Power	Number
kilo	10^3	1000
hecto	10^2	100
deka	10^1	10
————	10^0	1
deci	10^{-1}	.1
centi	10^{-2}	.01
milli	10^{-3}	.001
micro	10^{-6}	.000001

* RULES FOR ROUNDING OFF

Examples:

A. When the figures following the last number to be retained exceed 5, the last number retained is *increased by 1.*

$$2.8006 = 2.801$$
$$2.80052 = 2.801$$

B. When the figures following the last number to be retained is *less than 5,* the last number retained is *left* unchanged.

$$2.8003 = 2.800$$
$$2.80047 = 2.800$$

C. When the figures following the last number to be retained equal exactly 5, the last number retained is made the closest even value.

$$2.8015 = 2.802$$
$$2.8005 = 2.800$$

Using our conversion tables shown on pages 194 to 209 of this book, to convert 72 ± 0.1 or 2.835 ± .004 we find:

mm	inch		inch	mm
72	=	2.83464	2.835	=	72.0090
72	=	2.835	2.835	=	72.0
0.1	=	.00393	.004	=	0.1016
		(2.835 ± .004	=	(72 ± 0.1)

TO DETERMINE NUMBER OF DECIMAL PLACES TO RETAIN

Chart 1 — Millimeters to Inches*

Original Tolerance in Millimeter		Fineness of Rounding. Derived inch measure to be rounded to the nearest:
at least	less than	
0	±0.008	.000 01 (5 decimal places)
±0.008	±0.08	.000 1 (4 decimal places)
±0.08	±0.8	.001 (3 decimal places)
±0.8	±8	.01 (2 decimal places)
±8	and up	.1 (1 decimal place)

Chart 2 — Inches to Millimeters*

Original Tolerance in Inches		Fineness of Rounding. Derived millimeter measure to be rounded to the nearest:
at least	less than	
.000	±.000 16	0.000 1 (4 decimal places)
±.000 16	±.001 6	0.001 (3 decimal places)
±.001 6	±.016	0.01 (2 decimal places)
±.016	±.16	0.1 (1 decimal place)
±.16	and up	1 (0 decimal place)

Common Fractions of an Inch
to Millimeters

8ths	16ths	32nds	64ths	Decimal inch	Millimeters
			1	0.015625	0.396875
		1		.031250	0.793750
			3	.046875	1.190625
	1			.062500	1.587500
			5	.078125	1.984375
		3		0.093750	2.381250
			7	.109375	2.778125
1				.125000	3.175000
			9	.140625	3.571875
		5		.156250	3.968750
			11	0.171875	4.365625
	3			.187500	4.762500
			13	.203125	5.159375
		7		.218750	5.556250
			15	.234375	5.953125
				0.250000	6.350000
			17	.265625	6.746875
		9		.281250	7.143750
			19	.296875	7.540625
	5			.312500	7.937500
			21	0.328125	8.334375
		11		.343750	8.731250
			23	.359375	9.128125
3				.375000	9.525000
			25	.390625	9.921875
		13		0.406250	10.318750
			27	.421875	10.715625
	7			.437500	11.112500
			29	.453125	11.509375
		15		.468750	11.906250
			31	.484375	12.303125
				.500000	12.700000

NOTE: Table is exact; all figures beyond the six places given are zeros.

Common Fractions of an Inch to Millimeters

8ths	16ths	32nds	64ths	Decimal inch	Millimeters
			33	0 .515625	13 .096875
		17		.531250	13 .493750
			35	.546875	13 .890625
	9			.562500	14 .287500
			37	.578125	14 .684375
		19		0 .593750	15 .081250
			39	.609375	15 .478125
5				.625000	15 .875000
			41	.640625	16 .271875
		21		.656250	16 .668750
			43	0 .671875	17 .065625
	11			.687500	17 .462500
			45	.703125	17 .859375
		23		.718750	18 .256250
			47	.734375	18 .653125
				0 .750000	19 .050000
			49	.765625	19 .446875
		25		.781250	19 .843750
			51	.796875	20 .240625
	13			.812500	20 .637500
			53	0 .828125	21 .034375
		27		.843750	21 .431250
			55	.859375	21 .828125
7				.875000	22 .225000
			57	.890625	22 .621875
		29		0 .906250	23 .018750
			59	.921875	23 .415625
	15			.937500	23 .812500
			61	.953125	24 .209375
		31		.968750	24 .606250
			63	.984375	25 .003125
8	16	32	64	1 .000000	25 .400000

(Basis: 1 inch = 25.4 millimeters)

Inches to Millimeters
(Basis: 1 inch = 25.4 millimeters)

Inches	Millimeters	Inches	Millimeters
1	25.4	26	660.4
2	50.8	27	685.8
3	76.2	28	711.2
4	101.6	29	736.6
5	127.0	30	762.0
6	152.4	31	787.4
7	177.8	32	812.8
8	203.2	33	838.2
9	228.6	34	863.6
10	254.0	35	889.0
11	279.4	36	914.4
12	304.8	37	939.8
13	330.2	38	965.2
14	355.6	39	990.6
15	381.0	40	1016.0
16	406.4	41	1041.4
17	431.8	42	1066.8
18	457.2	43	1092.2
19	482.6	44	1117.6
20	508.0	45	1143.0
21	533.4	46	1168.4
22	558.8	47	1193.8
23	584.2	48	1219.2
24	609.6	49	1244.6
25	635.0	50	1270.0

NOTE: The above table is exact.

Inches to Millimeters
(Basis: 1 inch = 25.4 millimeters)

Inches	Millimeters	Inches	Millimeters
51	1295.4	76	1930.4
52	1320.8	77	1955.8
53	1346.2	78	1981.2
54	1371.6	79	2006.6
55	1397.0	80	2032.0
56	1422.4	81	2057.4
57	1447.8	82	2082.8
58	1473.2	83	2108.2
59	1498.6	84	2133.6
60	1524.0	85	2159.0
61	1549.4	86	2184.4
62	1574.8	87	2209.8
63	1600.2	88	2235.2
64	1625.6	89	2260.6
65	1651.0	90	2286.0
66	1676.4	91	2311.4
67	1701.8	92	2336.8
68	1727.2	93	2362.2
69	1752.6	94	2387.6
70	1778.0	95	2413.0
71	1803.4	96	2438.4
72	1828.8	97	2463.8
73	1854.2	98	2489.2
74	1879.6	99	2514.6
75	1905.0	100	2540.0

NOTE: The above table is exact.

Millimeters to Inches
(Basis: 1 inch = 25.4 millimeters)

Millimeters	Inches	Millimeters	Inches
1	0.039370	26	1.023622
2	.078740	27	1.062992
3	.118110	28	1.102362
4	.157480	29	1.141732
5	.196850	30	1.181102
6	.236220	31	1.220472
7	.275591	32	1.259843
8	.314961	33	1.299213
9	.354331	34	1.338583
10	.393701	35	1.377953
11	.433071	36	1.417323
12	.472441	37	1.456693
13	.511811	38	1.496063
14	.551181	39	1.535433
15	.590551	40	1.574803
16	.629921	41	1.614173
17	.669291	42	1.653543
18	.708661	43	1.692913
19	.748031	44	1.732283
20	.787402	45	1.771654
21	.826772	46	1.811024
22	.866142	47	1.850394
23	.905512	48	1.889764
24	.944882	49	1.929134
25	.984252	50	1.968504

NOTE: The above table is approximate:
$1/25.4 = 0.039370078740 +$.

Millimeters to Inches
(Basis: 1 inch = 25.4 millimeters)

Millimeters	Inches	Millimeters	Inches
51	2.007874	76	2.992126
52	2.047244	77	3.031496
53	2.086614	78	3.070866
54	2.125984	79	3.110236
55	2.165354	80	3.149606
56	2.204724	81	3.188976
57	2.244094	82	3.228346
58	2.283465	83	3.267717
59	2.322835	84	3.307087
60	2.362205	85	3.346457
61	2.401575	86	3.385827
62	2.440945	87	3.425197
63	2.480315	88	3.464567
64	2.519685	89	3.503937
65	2.559055	90	3.543307
66	2.598425	91	3.582677
67	2.637795	92	3.622047
68	2.677165	93	3.661417
69	2.716535	94	3.700787
70	2.755906	95	3.740157
71	2.795276	96	3.779528
72	2.834646	97	3.818898
73	2.874016	98	3.858268
74	2.913386	99	3.897638
75	2.952756	100	3.937008

NOTE: The above table is approximate:
1/25.4 = 0.039370078740 +.

CONVERSION CHART
INCH/MM

Drill No. or Letter		Inch	mm	Drill No. or Letter		Inch	mm
		.001	0.0254			.051	1.2954
		.002	0.0508	55		.052	1.3208
		.003	0.0762			.053	1.3462
		.004	0.1016			.054	1.3716
		.005	0.1270	54		.055	1.3970
		.006	0.1524			.056	1.4224
		.007	0.1778			.057	1.4478
		.008	0.2032			.058	1.4732
		.009	0.2286	53	.0595	.059	1.4986
		.010	0.2540			.060	1.5240
		.011	0.2794			.061	1.5494
		.012	0.3048			.062	1.5748
80	.0135	.013	0.3302		1/16	.0625	1.5875
79	.0145	.014	0.3556			.063	1.6002
		.015	0.3810	52	.0635	.064	1.6256
	1/64	.0156	0.3969			.065	1.6510
78		.016	0.4064			.066	1.6764
		.017	0.4318	51		.067	1.7018
77		.018	0.4572			.068	1.7272
		.019	0.4826			.069	1.7526
76		.020	0.5080	50		.070	1.7780
75		.021	0.5334			.071	1.8034
74	.0225	.022	0.5588			.072	1.8288
		.023	0.5842	49		.073	1.8542
73		.024	0.6096			.074	1.8796
72		.025	0.6350			.075	1.9050
71		.026	0.6604	48		.076	1.9304
		.027	0.6858			.077	1.9558
70		.028	0.7112	47	.0785	.078	1.9812
69	.0292	.029	0.7366		5/64	.0781	1.9844
		.030	0.7620			.0787	2.0000
68		.031	0.7874			.079	2.0066
	1/32	.0312	0.7937			.080	2.0320
67		.032	0.8128	46		.081	2.0574
66		.033	0.8382	45		.082	2.0828
		.034	0.8636			.083	2.1082
65		.035	0.8890			.084	2.1336
64		.036	0.9144			.085	2.1590
63		.037	0.9398	44		.086	2.1844
62		.038	0.9652			.087	2.2098
61		.039	0.9906			.088	2.2352
		.0394	1.0000	43		.089	2.2606
60		.040	1.0160			.090	2.2860
59		.041	1.0414			.091	2.3114
58		.042	1.0668			.092	2.3368
57		.043	1.0922	42	.0935	.093	2.3622
		.044	1.1176		3/32	.0937	2.3812
		.045	1.1430			.094	2.3876
56	.0465	.046	1.1684			.095	2.4130
	3/64	.0469	1.1906	41		.096	2.4384
		.047	1.1938			.097	2.4638
		.048	1.2192	40		.098	2.4892
		.049	1.2446			.099	2.5146
		.050	1.2700	39	.0995	.100	2.5400

CONVERSION CHART
INCH/MM

Drill No. or Letter		Inch	mm	Drill No. or Letter		Inch	mm
38	.1015	.101	2.5654	24		.152	3.8608
		.102	2.5908			.153	3.8862
		.103	2.6162	23		.154	3.9116
37		.104	2.6416			.155	3.9370
		.105	2.6670			.156	3.9624
36	.1065	.106	2.6924		5/32	.1562	3.9687
		.107	2.7178	22		.157	3.9878
		.108	2.7432			.1575	4.0000
		.109	2.7686			.158	4.0132
	7/64	.1094	2.7781	21		.159	4.0386
35		.110	2.7940			.160	4.0640
34		.111	2.8194	20		.161	4.0894
		.112	2.8448			.162	4.1148
33		.113	2.8702			.163	4.1402
		.114	2.8956			.164	4.1656
		.115	2.9210			.165	4.1910
32		.116	2.9464	19		.166	4.2164
		.117	2.9718			.167	4.2418
		.118	2.9972			.168	4.2672
		.1181	3.0000	18	.1635	.169	4.2926
		.119	3.0226			.170	4.3180
31		.120	3.0480			.171	4.3434
		.121	3.0734		11/64	.1719	4.3656
		.122	3.0988			.172	4.3688
		.123	3.1242	17		.173	4.3942
		.124	3.1496			.174	4.4196
	1/8	.125	3.1750			.175	4.4450
		.126	3.2004			.176	4.4704
		.127	3.2258	16		.177	4.4958
		.128	3.2512			.178	4.5212
30	.1285	.129	3.2766			.179	4.5466
		.130	3.3020	15		.180	4.5720
		.131	3.3274			.181	4.5974
		.132	3.3528	14		.182	4.6228
		.133	3.3782			.183	4.6482
		.134	3.4036			.184	4.6736
		.135	3.4290	13		.185	4.6990
29		.136	3.4544			.186	4.7244
		.137	3.4798			.187	4.7498
		.138	3.5052		3/16	.1875	4.7625
		.139	3.5306			.188	4.7752
28	.1405	.140	3.5560	12		.189	4.8006
	9/64	.1406	3.5719			.190	4.8260
		.141	3.5814	11		.191	4.8514
		.142	3.6068			.192	4.8768
		.143	3.6322			.193	4.9022
27		.144	3.6576	10	.1935	.194	4.9276
		.145	3.6830			.195	4.9530
		.146	3.7084	9		.196	4.9784
26		.147	3.7338			.1969	5.0000
		.148	3.7592			.197	5.0038
		.149	3.7846			.198	5.0292
25	.1495	.150	3.8100			.199	5.0546
		.151	3.8354	8		.200	5.0800

CONVERSION CHART
INCH/MM

Drill No. or Letter		Inch	mm	Drill No. or Letter		Inch	mm
7		.201	5.1054			.251	6.3754
		.202	5.1308			.252	6.4008
		.203	5.1562			.253	6.4262
	13/64	.2031	5.1594			.254	6.4516
6		.204	5.1816			.255	6.4770
5	.2055	.205	5.2070			.256	6.5024
		.206	5.2324	F		.257	6.5278
		.207	5.2578			.258	6.5532
		.208	5.2832			.259	6.5786
4		.209	5.3086			.260	6.6040
		.210	5.3340	G		.261	6.6294
		.211	5.3594			.262	6.6548
		.212	5.3848			.263	6.6802
3		.213	5.4102			.264	6.7056
		.214	5.4356			.265	6.7310
		.215	5.4610		17/64	.2656	6.7469
		.216	5.4864	H		.266	6.7564
		.217	5.5118			.267	6.7818
		.218	5.5372			.268	6.8072
	7/32	.2187	5.5562			.269	6.8326
		.219	5.5626			.270	6.8580
		.220	5.5880			.271	6.8834
2		.221	5.6134	I		.272	6.9088
		.222	5.6388			.273	6.9342
		.223	5.6642			.274	6.9596
		.224	5.6896			.275	6.9850
		.225	5.7150			.2756	7.0000
		.226	5.7404			.276	7.0104
		.227	5.7658	J		.277	7.0358
1		.228	5.7912			.278	7.0612
		.229	5.8166			.279	7.0866
		.230	5.8420			.280	7.1120
		.231	5.8674	K		.281	7.1374
		.232	5.8928		9/32	.2812	7.1437
		.233	5.9182			.282	7.1628
A		.234	5.9436			.283	7.1882
	15/64	.2344	5.9531			.284	7.2136
		.235	5.9690			.285	7.2390
		.236	5.9944			.286	7.2644
		.2362	6.0000			.287	7.2898
		.237	6.0198			.288	7.3152
B		.238	6.0452			.289	7.3406
		.239	6.0706			.290	7.3660
		.240	6.0960	L		.291	7.3914
		.241	6.1214			.292	7.4168
C		.242	6.1468			.293	7.4422
		.243	6.1722			.294	7.4676
		.244	6.1976	M		.295	7.4930
		.245	6.2230			.296	7.5184
D		.246	6.2484		19/64	.2969	7.5406
		.247	6.2738			.297	7.5438
		.248	6.2992			.298	7.5692
		.249	6.3246			.299	7.5946
E	1/4	.250	6.3500			.300	7.6200

CONVERSION CHART
INCH/MM

Drill No. or Letter	Inch	mm	Drill No. or Letter	Inch	mm
N	.301	7.6454		.351	8.9154
	.302	7.6708		.352	8.9408
	.303	7.6962		.353	8.9662
	.304	7.7216		.354	8.9916
	.305	7.7470		.3543	9.0000
	.306	7.7724		.355	9.0170
	.307	7.7978		.356	9.0424
	.308	7.8232		.357	9.0678
	.309	7.8486	T	.358	9.0932
	.310	7.8740		.359	9.1185
	.311	7.8994	23/64	.3594	9.1281
	.312	7.9248		.360	9.1440
5/16	.3125	7.9375		.361	9.1694
	.313	7.9502		.362	9.1948
	.314	7.9756		.363	9.2202
	.3150	8.0000		.364	9.2456
	.315	8.0010		.365	9.2710
O	.316	8.0264		.366	9.2964
	.317	8.0518		.367	9.3218
	.318	8.0772	U	.368	9.3472
	.319	8.1026		.369	9.3726
	.320	8.1280		.370	9.3980
	.321	8.1534		.371	9.4234
	.322	8.1788		.372	9.4488
P	.323	8.2042		.373	9.4742
	.324	8.2296		.374	9.4996
	.325	8.2550	3/8	.375	9.5250
	.326	8.2804		.376	9.5504
	.327	8.3058	V	.377	9.5758
	.328	8.3312		.378	9.6012
21/64	.3281	8.3344		.379	9.6266
	.329	8.3566		.380	9.6520
	.330	8.3820		.381	9.6774
	.331	8.4074		.382	9.7028
Q	.332	8.4328		.383	9.7282
	.333	8.4582		.384	9.7536
	.334	8.4836		.385	9.7790
	.335	8.5090	W	.386	9.8044
	.336	8.5344		.387	9.8298
	.337	8.5598		.388	9.8552
	.338	8.5852		.389	9.8806
R	.339	8.6106		.390	9.9060
	.340	8.6360	25/64	.3906	9.9219
	.341	8.6614		.391	9.9314
	.342	8.6868		.392	9.9568
	.343	8.7122		.393	9.9822
11/32	.3437	8.7312		.3937	10.0000
	.344	8.7376		.394	10.0076
	.345	8.7630		.395	10.0330
	.346	8.7884		.396	10.0584
	.347	8.8138	X	.397	10.0838
S	.348	8.8392		.398	10.1092
	.349	8.8646		.399	10.1346
	.350	8.8900		.400	10.1600

CONVERSION CHART INCH/MM

Drill No. or Letter	Inch	mm	Drill No. or Letter	Inch	mm
	.401	10.1854		.451	11.4554
	.402	10.2108		.452	11.4808
	.403	10.2362		.453	11.5062
Y	.404	10.2616	29/64	.4531	11.5094
	.405	10.2870		.454	11.5316
	.406	10.3124		.455	11.5570
13/32	.4062	10.3187		.456	11.5824
	.407	10.3378		.457	11.6078
	.408	10.3632		.458	11.6332
	.409	10.3886		.459	11.6586
	.410	10.4140		.460	11.6840
	.411	10.4394		.461	11.7094
	.412	10.4648		.462	11.7348
Z	.413	10.4902		.463	11.7602
	.414	10.5156		.464	11.7856
	.415	10.5410		.465	11.8110
	.416	10.5664		.466	11.8364
	.417	10.5918		.467	11.8618
	.418	10.6172		.468	11.8872
	.419	10.6426	15/32	.4687	11.9062
	.420	10.6680		.469	11.9126
	.421	10.6934		.470	11.9380
27/64	.4219	10.7156		.471	11.9634
	.422	10.7188		.472	11.9888
	.423	10.7442		.4724	12.0000
	.424	10.7696		.473	12.0142
	.425	10.7950		.474	12.0396
	.426	10.8204		.475	12.0650
	.427	10.8458		.476	12.0904
	.428	10.8712		.477	12.1158
	.429	10.8966		.478	12.1412
	.430	10.9220		.479	12.1666
	.431	10.9474		.480	12.1920
	.432	10.9728		.481	12.2174
	.433	10.9982		.482	12.2428
	.4331	11.0000		.483	12.2682
	.434	11.0236		.484	12.2936
	.435	11.0490	31/64	.4844	12.3031
	.436	11.0744		.485	12.3190
	.437	11.0998		.486	12.3444
7/16	.4375	11.1125		.487	12.3698
	.438	11.1252		.488	12.3952
	.439	11.1506		.489	12.4206
	.440	11.1760		.490	12.4460
	.441	11.2014		.491	12.4714
	.442	11.2268		.492	12.4968
	.443	11.2522		.493	12.5222
	.444	11.2776		.494	12.5476
	.445	11.3030		.495	12.5730
	.446	11.3284		.496	12.5984
	.447	11.3538		.497	12.6238
	.448	11.3792		.498	12.6492
	.449	11.4046		.499	12.6746
	.450	11.4300	1/2	.500	12.7000

CONVERSION CHART INCH/MM

Inch		mm	Inch		mm
	.501	12.7254		.551	13.9954
	.502	12.7508		.5512	14.0000
	.503	12.7762		.552	14.0208
	.504	12.8016		.553	14.0462
	.505	12.8270		.554	14.0716
	.506	12.8524		.555	14.0970
	.507	12.8778		.556	14.1224
	.508	12.9032		.557	14.1478
	.509	12.9286		.558	14.1732
	.510	12.9540		.559	14.1986
	.511	12.9794		.560	14.2240
	.5118	13.0000		.561	14.2494
	.512	13.0048		.562	14.2748
	.513	13.0302	9/16	.5625	14.2875
	.514	13.0556		.563	14.3002
	.515	13.0810		.564	14.3256
33/64	.5156	13.0968		.565	14.3510
	.516	13.1064		.566	14.3764
	.517	13.1318		.567	14.4018
	.518	13.1572		.568	14.4272
	.519	13.1826		.569	14.4526
	.520	13.2080		.570	14.4780
	.521	13.2334		.571	14.5034
	.522	13.2588		.572	14.5288
	.523	13.2842		.573	14.5542
	.524	13.3096		.574	14.5796
	.525	13.3350		.575	14.6050
	.526	13.3604		.576	14.6304
	.527	13.3858		.577	14.6558
	.528	13.4112		.578	14.6812
	.529	13.4366	37/64	.5781	14.6844
	.530	13.4620		.579	14.7066
	.531	13.4874		.580	14.7320
17/32	.5312	13.4937		.581	14.7574
	.532	13.5128		.582	14.7828
	.533	13.5382		.583	14.8082
	.534	13.5636		.584	14.8336
	.535	13.5890		.585	14.8590
	.536	13.6144		.586	14.8844
	.537	13.6398		.587	14.9098
	.538	13.6652		.588	14.9352
	.539	13.6906		.589	14.9606
	.540	13.7160		.590	14.9860
	.541	13.7414		.5906	15.0000
	.542	13.7668		.591	15.0114
	.543	13.7922		.592	15.0368
	.544	13.8176		.593	15.0622
	.545	13.8430	19/32	.5937	15.0812
	.546	13.8684		.594	15.0876
35/64	.5469	13.8906		.595	15.1130
	.547	13.8938		.596	15.1384
	.548	13.9192		.597	15.1638
	.549	13.9446		.598	15.1892
	.550	13.9700		.599	15.2146

CONVERSION CHART
INCH/MM

Inch		mm	Inch		mm
	.600	15.2400		.651	16.5354
	.601	15.2654		.652	16.5608
	.602	15.2908		.653	16.5862
	.603	15.3162		.654	16.6116
	.604	15.3416		.655	16.6370
	.605	15.3670		.656	16.6624
	.606	15.3924	21/32	.6562	16.6687
	.607	15.4178		.657	16.6878
	.608	15.4432		.658	16.7132
	.609	15.4686		.659	16.7386
39/64	.6094	15.4781		.660	16.7640
	.610	15.4940		.661	16.7894
	.611	15.5194		.662	16.8148
	.612	15.5448		.663	16.8402
	.613	15.5702		.664	16.8656
	.614	15.5956		.665	16.8910
	.615	15.6210		.666	16.9164
	.616	15.6464		.667	16.9418
	.617	15.6718		.668	16.9672
	.618	15.6972		.669	16.9926
	.619	15.7226		.6693	17.0000
	.620	15.7480		.670	17.0180
	.621	15.7734		.671	17.0434
	.622	15.7988	43/64	.6719	17.0656
	.623	15.8242		.672	17.0688
	.624	15.8496		.673	17.0942
5/8	.625	15.8750		.674	17.1196
	.626	15.9004		.675	17.1450
	.627	15.9258		.676	17.1704
	.628	15.9512		.677	17.1958
	.629	15.9766		.678	17.2212
	.6299	16.0000		.679	17.2466
	.630	16.0020		.680	17.2720
	.631	16.0274		.681	17.2974
	.632	16.0528		.682	17.3228
	.633	16.0782		.683	17.3482
	.634	16.1036		.684	17.3736
	.635	16.1290		.685	17.3990
	.636	16.1544		.686	17.4244
	.637	16.1798		.687	17.4498
	.638	16.2052	11/16	.6875	17.4625
	.639	16.2306		.688	17.4752
	.640	16.2560		.689	17.5006
41/64	.6406	16.2719		.690	17.5260
	.641	16.2814		.691	17.5514
	.642	16.3068		.692	17.5768
	.643	16.3322		.693	17.6022
	.644	16.3576		.694	17.6276
	.645	16.3830		.695	17.6530
	.646	16.4084		.696	17.6784
	.647	16.4338		.697	17.7038
	.648	16.4592		.698	17.7292
	.649	16.4846		.699	17.7546
	.650	16.5100		.700	17.7800

CONVERSION CHART INCH/MM

Inch		mm	Inch		mm
	.701	17.8054		.751	19.0754
	.702	17.8308		.752	19.1008
	.703	17.8562		.753	19.1262
45/64	.7031	17.8594		.754	19.1516
	.704	17.8816		.755	19.1770
	.705	17.9070		.756	19.2024
	.706	17.9324		.757	19.2278
	.707	17.9578		.758	19.2532
	.708	17.9832		.759	19.2786
	.7087	18.0000		.760	19.3040
	.709	18.0086		.761	19.3294
	.710	18.0340		.762	19.3548
	.711	18.0594		.763	19.3802
	.712	18.0848		.764	19.4056
	.713	18.1102		.765	19.4310
	.714	18.1356	49/64	.7656	19.4469
	.715	18.1610		.766	19.4564
	.716	18.1864		.767	19.4818
	.717	18.2118		.768	19.5072
	.718	18.2372		.769	19.5326
23/32	.7187	18.2562		.770	19.5580
	.719	18.2626		.771	19.5834
	.720	18.2880		.772	19.6088
	.721	18.3134		.773	19.6342
	.722	18.3388		.774	19.6596
	.723	18.3642		.775	19.6850
	.724	18.3896		.776	19.7104
	.725	18.4150		.777	19.7358
	.726	18.4404		.778	19.7612
	.727	18.4658		.779	19.7866
	.728	18.4912		.780	19.8120
	.729	18.5166		.781	19.8374
	.730	18.5420	25/32	.7812	19.8433
	.731	18.5674		.782	19.8628
	.732	18.5928		.783	19.8882
	.733	18.6182		.784	19.9136
	.734	18.6436		.785	19.9390
47/64	.7344	18.6532		.786	19.9644
	.735	18.6690		.787	19.9898
	.736	18.6944		.7874	20.0000
	.737	18.7198		.788	20.0152
	.738	18.7452		.789	20.0406
	.739	18.7706		.790	20.0660
	.740	18.7960		.791	20.0914
	.741	18.8214		.792	20.1168
	.742	18.8468		.793	20.1422
	.743	18.8722		.794	20.1676
	.744	18.8976		.795	20.1930
	.745	18.9230		.796	20.2184
	.746	18.9484	51/64	.7969	20.2402
	.747	18.9738		.797	20.2438
	.748	18.9992		.798	20.2692
	.7480	19.0000		.799	20.2946
	.749	19.0246			
3/4	.750	19.0500			

CONVERSION CHART
INCH/MM

Inch		mm	Inch		mm
	.800	20.3200		.851	21.6154
	.801	20.3454		.852	21.6408
	.802	20.3708		.853	21.6662
	.803	20.3962		.854	21.6916
	.804	20.4216		.855	21.7170
	.805	20.4470		.856	21.7424
	.806	20.4724		.857	21.7678
	.807	20.4978		.858	21.7932
	.808	20.5232		.859	21.8186
	.809	20.5486	55/64	.8594	21.8281
	.810	20.5740		.860	21.8440
	.811	20.5994		.861	21.8694
	.812	20.6248		.862	21.8948
13/16	.8125	20.6375		.863	21.9202
	.813	20.6502		.864	21.9456
	.814	20.6756		.865	21.9710
	.815	20.7010		.866	21.9964
	.816	20.7264		.8661	22.0000
	.817	20.7518		.867	22.0218
	.818	20.7772		.868	22.0472
	.819	20.8026		.869	22.0726
	.820	20.8280		.870	22.0980
	.821	20.8534		.871	22.1234
	.822	20.8788		.872	22.1488
	.823	20.9042		.873	22.1742
	.824	20.9296		.874	22.1996
	.825	20.9550	7/8	.875	22.2250
	.826	20.9804		.876	22.2504
	.8268	21.0000		.877	22.2758
	.827	21.0058		.878	22.3012
	.828	21.0312		.879	22.3266
53/64	.8281	21.0344		.880	22.3520
	.829	21.0566		.881	22.3774
	.830	21.0820		.882	22.4028
	.831	21.1074		.883	22.4282
	.832	21.1328		.884	22.4536
	.833	21.1582		.885	22.4790
	.834	21.1836		.886	22.5044
	.835	21.2090		.887	22.5298
	.836	21.2344		.888	22.5552
	.837	21.2598		.889	22.5806
	.838	21.2852		.890	22.6060
	.839	21.3106	57/64	.8906	22.6219
	.840	21.3360		.891	22.6314
	.841	21.3614		.892	22.6568
	.842	21.3868		.893	22.6822
	.843	21.4122		.894	22.7076
27/32	.8437	21.4312		.895	22.7330
	.844	21.4376		.896	22.7584
	.845	21.4630		.897	22.7838
	.846	21.4884		.898	22.8092
	.847	21.5138		.899	22.8346
	.848	21.5392		.900	22.8600
	.849	21.5646			
	.850	21.5900			

CONVERSION CHART
INCH/MM

Inch		mm	Inch		mm
	.901	22.8854		.951	24.1554
	.902	22.9108		.952	24.1808
	.903	22.9362		.953	24.2062
	.904	22.9616	61/64	.9531	24.2094
	.905	22.9870		.954	24.2316
	.9055	23.0000		.955	24.2570
	.906	23.0124		.956	24.2824
29/32	.9062	23.0187		.957	24.3078
	.907	23.0378		.958	24.3332
	.908	23.0632		.959	24.3586
	.909	23.0886		.960	24.3840
	.910	23.1140		.961	24.4094
	.911	23.1394		.962	24.4348
	.912	23.1648		.963	24.4602
	.913	23.1902		.964	24.4856
	.914	23.2156		.965	24.5110
	.915	23.2410		.966	24.5364
	.916	23.2664		.967	24.5618
	.917	23.2918		.968	24.5872
	.918	23.3172	31/32	.9687	24.6062
	.919	23.3426		.969	24.6126
	.920	23.3680		.970	24.6380
	.921	23.3934		.971	24.6634
59/64	.9219	23.4156		.972	24.6888
	.922	23.4188		.973	24.7142
	.923	23.4442		.974	24.7396
	.924	23.4696		.975	24.7650
	.925	23.4950		.976	24.7904
	.926	23.5204		.977	24.8158
	.927	23.5458		.978	24.8412
	.928	23.5712		.979	24.8666
	.929	23.5966		.980	24.8920
	.930	23.6220		.981	24.9174
	.931	23.6474		.982	24.9428
	.932	23.6728		.983	24.9682
	.933	23.6982		.984	24.9936
	.934	23.7236		.9843	25.0000
	.935	23.7490	63/64	.9844	25.0031
	.936	23.7744		.985	25.0190
	.937	23.7998		.986	25.0444
15/16	.9375	23.8125		.987	25.0698
	.938	23.8252		.988	25.0952
	.939	23.8506		.989	25.1206
	.940	23.8760		.990	25.1460
	.941	23.9014		.991	25.1714
	.942	23.9268		.992	25.1968
	.943	23.9522		.993	25.2222
	.944	23.9776		.994	25.2476
	.9449	24.0000		.995	25.2730
	.945	24.0030		.996	25.2984
	.946	24.0284		.997	25.3238
	.947	24.0538		.998	25.3492
	.948	24.0792		.999	25.3746
	.949	24.1046		1.000	25.4000
	.950	24.1300			

A COMPARISON OF THERMOMETRIC SCALES
Kelvin; Celsius; Fahrenheit

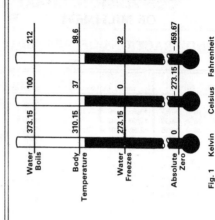

Fig. 1   Kelvin   Celsius   Fahrenheit

Definitions:

Kelvin (K) — The kelvin scale has its origin or zero point at absolute zero, and the freezing point of water is defined as 273.15 kelvins. Graduations on the kelvin scale are equal to those on the Celsius scale.

Celsius (°C) — The Celsius scale uses the freezing point of water as its zero point, and the boiling point of water as 100 degrees. Thus the interval between these two points is graduated into 100 divisions.

Fahrenheit (F) — The Fahrenheit scale uses the freezing point of water as 32 degrees, and the boiling point of water as 212 degrees. Thus the interval between these points is graduated into 180 divisions.

Note: See Fig. 1 for actual comparisons.

Conversion formulas:

$$K = °C + 273.15$$

$$°C = (F - 32) \times 5/9; \text{ or } °C = \frac{(F - 32)}{1.8}$$

$$F = (°C \times 9/5) + 32; \text{ or } F = 1.8 °C + 32$$

DECIMAL EQUIVALENTS
OF MILLIMETERS
AND FRACTIONS OF MILLIMETERS

1/100 mm = .01 mm = .0003937 inches

mm	Inches	mm	Inches	mm	Inches	mm	Inches
1/50 = .02 = .00079		39/50 = .78 = .03071		27 = 1.06299		64 = 2.51968	
2/50 = .04 = .00157		40/50 = .80 = .03150		28 = 1.10236		65 = 2.55905	
3/50 = .06 = .00236		41/50 = .82 = .03228		29 = 1.14173		66 = 2.59842	
4/50 = .08 = .00315		42/50 = .84 = .03307		30 = 1.18110		67 = 2.63779	
5/50 = .10 = .00394		43/50 = .86 = .03386		31 = 1.22047		68 = 2.67716	
6/50 = .12 = .00472		44/50 = .88 = .03465		32 = 1.25984		69 = 2.71653	
7/50 = .14 = .00551		45/50 = .90 = .03543		33 = 1.29921		70 = 2.75590	
8/50 = .16 = .00630		46/50 = .92 = .03622		34 = 1.33858		71 = 2.79527	
9/50 = .18 = .00709		47/50 = .94 = .03701		35 = 1.37795		72 = 2.83464	
10/50 = .20 = .00787		48/50 = .96 = .03780		36 = 1.41732		73 = 2.87401	
11/50 = .22 = .00866		49/50 = .98 = .03858		37 = 1.45669		74 = 2.91338	
12/50 = .24 = .00945		1 = .03937		38 = 1.49606		75 = 2.95275	
13/50 = .26 = .01024		2 = .07874		39 = 1.53543		76 = 2.99212	
14/50 = .28 = .01102		3 = .11811		40 = 1.57480		77 = 3.03149	
15/50 = .30 = .01181		4 = .15748		41 = 1.61417		78 = 3.07086	
16/50 = .32 = .01260		5 = .19685		42 = 1.65354		79 = 3.11023	
17/50 = .34 = .01339		6 = .23622		43 = 1.69291		80 = 3.14960	
18/50 = .36 = .01417		7 = .27559		44 = 1.73228		81 = 3.18897	
19/50 = .38 = .01496		8 = .31496		45 = 1.77165		82 = 3.22834	
20/50 = .40 = .01575		9 = .35433		46 = 1.81102		83 = 3.26771	
21/50 = .42 = .01654		10 = .39370		47 = 1.85039		84 = 3.30708	
22/50 = .44 = .01732		11 = .43307		48 = 1.88976		85 = 3.34645	
23/50 = .46 = .01811		12 = .47244		49 = 1.92913		86 = 3.38582	
24/50 = .48 = .01890		13 = .51181		50 = 1.96850		87 = 3.42519	
25/50 = .50 = .01969		14 = .55118		51 = 2.00787		88 = 3.46456	
26/50 = .52 = .02047		15 = .59055		52 = 2.04724		89 = 3.50393	
27/50 = .54 = .02126		16 = .62992		53 = 2.08661		90 = 3.54330	
28/50 = .56 = .02205		17 = .66929		54 = 2.12598		91 = 3.58267	
29/50 = .58 = .02283		18 = .70866		55 = 2.16535		92 = 3.62204	
30/50 = .60 = .02362		19 = .74803		56 = 2.20472		93 = 3.66141	
31/50 = .62 = .02441		20 = .78740		57 = 2.24409		94 = 3.70078	
32/50 = .64 = .02520		21 = .82677		58 = 2.28346		95 = 3.74015	
33/50 = .66 = .02598		22 = .86614		59 = 2.32283		96 = 3.77952	
34/50 = .68 = .02677		23 = .90551		60 = 2.36220		97 = 3.81889	
35/50 = .70 = .02756		24 = .94488		61 = 2.40157		98 = 3.85826	
36/50 = .72 = .02835		25 = .98425		62 = 2.44094		99 = 3.89763	
37/50 = .74 = .02913		26 = 1.02362		63 = 2.48031		100 = 3.93700	
38/50 = .76 = .02992							

10 mm = 1 centimeter	=	0.3937 inches
10 cm = 1 decimeter	=	3.937 inches
10 dm = 1 meter	=	39.37 inches
25.4 mm = 1 inch		

READING METRIC MICROMETERS

READING TO 0.01mm

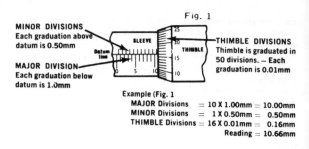

Fig. 1

MINOR DIVISIONS
Each graduation above datum is 0.50mm

MAJOR DIVISION
Each graduation below datum is 1.0mm

THIMBLE DIVISIONS
Thimble is graduated in 50 divisions. — Each graduation is 0.01mm

Example (Fig. 1
MAJOR Divisions = 10 X 1.00mm = 10.00mm
MINOR Divisions = 1 X 0.50mm = 0.50mm
THIMBLE Divisions = 16 X 0.01mm = 0.16mm
Reading = 10.66mm

READING TO 0.002mm

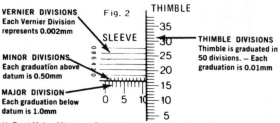

Fig. 2

THIMBLE

VERNIER DIVISIONS
Each Vernier Division represents 0.002mm

MINOR DIVISIONS
Each graduation above datum is 0.50mm

MAJOR DIVISION
Each graduation below datum is 1.0mm

THIMBLE DIVISIONS
Thimble is graduated in 50 divisions. — Each graduation is 0.01mm

1) Read Major, Minor and Thimble Divisions as described in Figure 1.

2) Note which Vernier graduation coincides with a line on the Thimble and read the micrometer as follows:

Example (Fig. 2)
MAJOR Divisions = 10 X 1.00mm = 10.00mm
MINOR Divisions = 1 X 0.50mm = 0.50mm
THIMBLE Divisions = 16 X 0.01mm = 0.16mm
*VERNIER Divisions = 2 X 0.002mm = 0.006mm
Reading = 10.666mm

*VERNIER line coincident with line on THIMBLE in the example is the 3rd Division. Since each Vernier line has a value of 0.002mm, this line is marked with the numeral 6 to permit reading the value directly.

Note: If Vernier line marked 0 is coincident with a line on the Thimble, the reading is an exact one-hundredth of a millimetre taken from the Thimble line which is coincident with the Datum line.

HOW TO READ VERNIER
with
ENGLISH/METRIC GRADUATIONS

METRIC READING – EXTERNAL

Use the upper Vernier and Beam Scales.

Each beam graduation is 1mm; Numbers on Beam show each 10mm; Vernier graduations are equal to 0.02mm and are numbered to show each 0.10mm.

Example in photo: 70 plus 8 Beam graduations = 78mm plus Vernier reading of coincidental Beam and Vernier graduations = 0.08mm.

Total External reading: 78 + 0.08 = 78.08mm

For Internal Measurement, read as above and add 7.62mm (width of Nibs) to the reading.

Example: 78.08 + 7.62 = 85.70mm

Reading = 78.08mm

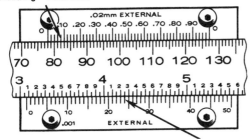

Reading = 3.074"

ENGLISH READING – EXTERNAL

Use the lower Vernier and Beam Scales.

Each Beam graduation is .050"; each numbered Beam graduation between inches is .100"; Vernier shows each .001" and has numbered graduations by .010".

Example in photo: 3.00" + .050" (Beam Reading) + .024" (Vernier Reading) = 3.074".

For Internal Measurement, read as above and add .300" (width of Nibs) to the reading.

Example: 3.074" + .300" = 3.374'

213

NOTES

INDEX

NOTES